ANDREW
YOUNGEST LUMBERJACK

by

Dr. Duane R. Lund

Distributed by
Adventure Publications
P.O. Box 269
Cambridge, Minnesota 55008

ISBN 0-934860-62-9.

Andrew, Youngest Lumberjack

Printed by
Nordell Graphic Communications, Inc
Staples, Minnesota 56479

Printed in the United States of America

dedicated to my grandfather, Andrew Anderson

TABLE OF CONTENTS

CHAPTER I

"IT'S AN ILL WIND THAT BLOWS NO GOOD"

As long as he would live, Andrew would not forget the sights or the sounds of that moment. His father's cries of "Help!" came from the pasture just behind the barn. As Andrew came around the corner of the milk shed at a full run, he saw his father on the ground, lying just on the other side of the pasture gate. Torger, the bull, was only a few feet away, bellowing and pawing the earth. Pieces of sod and dirt filled the air behind the animal. A pitchfork was leaning against the wall of the barn next to the door, and Andrew grabbed it as he ran by without breaking stride. With no thought of his own safety, he vaulted over the barbed wire fence, a feat he later found himself totally incapable of doing, and ran between his fallen father and the angry bull. Again and again he jabbed the animal in the face with the sharp tines of the pitchfork until old Torger had enough and turned away. As Andrew looked over his shoulder, he was relieved to see that his father had dragged himself under the fence to safety. Andrew dropped the pitchfork and dove between the strands of wire to join his father. Actually, he need not have hurried. Torger had had enough and had turned his attention to one of the dairy cows (Hazel by name). He had been trying to mate with Hazel when he had noticed Mr. Anderson in the pasture and decided he might be a rival for the cow's affections. When the bull charged, the farmer had run for the gate and was just climbing over when the bull slammed into him, pinning him against the wooden structure.

Andrew discovered his knees were so weak and shaky from all the excitement that he had to grab hold of a fence post to pull himself to his feet. He then turned his attention to his father who was still lying on the ground and, as he caught his breath, asked, "Are you all right, Pa?"

"Not really," Mr. Anderson replied in his heavy Swedish accent. "I'm afraid my leg is broken."

And indeed it was, badly, in two places.

"Just stay put, Pa," Andrew urged. "I'll get some help."

Andrew ran towards the house, yelling, "Ma! Ma!"

His mother had not heard the commotion when the bull attacked, but she did hear her son's piercing cries and, knowing something was terribly wrong, swung open the kitchen door before Andrew was halfway there. As the boy ran up, his jumbled words poured out in segments, in between gasps for breath, "Pa's hurt, broken leg, Torger got him!"

Andrew's slightly older sister, Louise, had also heard the yelling and appeared in the doorway just behind her mother.

The stricken farmer was not visible from the house, but the mother and daughter quickly followed Andrew out to the back of the barn where Mr. Anderson had indeed been content to "stay put" and was carefully feeling of his right leg.

After assurances that the injuries were limited to his leg, the farmer was allowed to tell the brief but scary story of what had happened and how Torger, the trusted bull which had never threatened anyone, had attacked without warning.

Mrs. Anderson took charge and carefully wrapped pieces of thin but strong wooden lath on either side of the leg to prevent further damage. She then directed the making of a litter from a blanket and two poles so that her husband could be carried into the house and made more comfortable in the downstairs bedroom. Only then was there thought of fetching a doctor.

Doc Will was officed in Little Falls, the only community in the area big enough to support a physician, and it became Andrew's assignment to saddle up Dan, one of the horses, and head for town.

Although time was no longer critical, Andrew urged the horse to a full gallop down most of the 15 miles of dirt road to Little Falls.

Doc Will's office was in his home, and Andrew knew where that was because of his visits there a couple of years earlier when he had broken his arm in a fall from the hay mow. Luckily, the physician was in and had only one patient waiting to see him. As soon as he had been assured that the doctor would come right away, Andrew remounted Dan and headed home to tell the anxious family that help was coming.

It seemed forever, but the physician arrived in his buggy less than a half hour behind Andrew.

Doc Will proclaimed the injury a "nasty break, both the tibia and the fibula."

With skill and care, the doctor set the leg. The pain was great, but Mr. Anderson bravely gritted his teeth and took it, the sweat literally pouring off his brow. Splints were again bound in place and the patient told, "Move as little as possible."

"But Dr. Will," the farmer implored, his voice betraying deep concern, "How long before I can walk? How long before I can work?"

"Oh, a good six weeks before you can take off the splints, maybe a little longer. Meanwhile you'll need crutches. Maybe Andrew can make you a pair," came the reply.

"Now, Anders," Mrs. Anderson addressed her husband in an accent indicating that she was also born in Sweden, "I know what you're thinking, but don't you worry. Louise and Andrew and me, we'll do the farming; it will be just a little more effort for each one of us."

"Ya, Anna," Mr. Anderson replied to his wife, "but it's not just the farming I worry about. In just a week or so I'm supposed to leave for logging camp, you know."

At this point, Dr. Will interrupted, "Anders, there's just no way you're going to log this winter. It's far more important that your leg heal well and heal straight or your years of working will be severely limited."

"But, Dr. Will, you don't understand," Mr. Anderson protested. "With the money I save from logging we buy more land each year so that when Andrew grows up he'll be able to make a living here, too."

"Now listen here, Anders," the doctor argued, "missing one year is not going to make that much difference. It is far more important that your leg heal right or you may never return to the logging camp!"

"But we don't have that much time," Mr. Anderson explained. "Andrew is already fifteen years old; he'll soon be a man. The first thing you know, he'll get married and have a family and what we have going here just can't support two families. Besides, there's just forty acres of land left that our neighbor, Gust Kuhlander, wants to sell, and either Jensens or Olsons could grab it up whenever they get a little money ahead."

"But, Pa," Andrew consoled, "one year won't make all that much difference, and I don't plan *ever* to get married!"

This brought laughter from Mrs. Anderson and Louise, as the latter asked, "You want to bet?"

"You bet I want to bet!" Andrew snorted.

"Now, children," Mrs. Anderson interrupted, "there'll be no betting."

Then turning to her stricken husband, she added, "Anyway, Anders, it's settled, you'll not be going to logging camp this winter and Andrew and Louise and me, we'll do the chores. We're just lucky it's fall and the harvest is in. About all we got to do is feed the livestock and chickens and milk the cows. And remember," she added, brightening, "Andrew isn't going to school this year; he'll be home to help."

"That's another thing," Mr. Anderson continued his argument, "Andrew should have gone to high school in Little Falls this year. An eighth grade education isn't good enough anymore."

At this point Dr. Will stood up, put on his coat, and said as he moved toward the bedroom door, "Well, I'm not going to get into a family argument. I'll be out to see you in a week, Anders. If anything seems wrong with your leg, send Andrew and I'll be right out."

"Wait, Doctor," Mrs. Anderson also got up, "I must pay you."

"Well, give me a dollar if you have it, and drop off a couple of chickens when you come to town."

"Ya, sure, "Mrs. Anderson replied. "We really are thankful that you came so soon."

As the doctor left, Mrs. Anderson suggested that they let the patient get some rest. As each member of the family took their leave to go about their separate chores, each took stock of the big change the accident could make in their lives.

As Andrew pondered his father's misfortune, it suddenly occurred to him that the family tragedy might present a golden opportunity, for himself. This could be his big chance to go to logging camp, in his father's place! He knew he was too young to be a lumberjack, but there were other jobs, like helping the cook or taking care of the horses. The more he thought about it, the more enthusiastic he became. "And if there should be a shortage of lumberjacks," he told himself, "I have had a lot of experience helping clear land." Thoughtfully, he concluded, "It's like Pa always says, 'It's an ill wind that blows no good!' "

Andrew knew his parents would be opposed to his going to logging camp, mostly because of his age and because they wanted him to continue his education rather than get in the habit of going logging every winter. So he practiced his arguments thoroughly before he brought up the subject. He even wrote down all the reasons why he should be allowed to go to logging camp, then listed all the arguments he thought his folks might put up and then thought up answers for each one of them.

Andrew waited patiently for just the right time, but it was really hard to wait; he wanted to go so badly. He had never even been out of Morrison County, except once to visit his aunt and uncle down in St. Cloud. Finally, he could contain himself no longer. The occasion he chose was five days after the accident, one evening when his father had been served one of his favorite suppers, pork roast with lots of gravy over boiled potatoes with sweetened baby carrots, topped off with a big wedge of chocolate devil's food cake. As his mother entered the bedroom to pick up the dishes, Andrew followed her in and asked if she would stay while he asked an important question.

Without waiting for her response, Andrew began the little speech he had rehearsed so many times, "Ma, Pa, I've really got a great idea. I could take Pa's place in logging camp."

Andrew hadn't meant to let his folks get a word in until he had shared the whole plan, but his nervousness made him stop to catch his breath. Both spoke at the same time. His mother said, "I should say not; out of the question." His father simply said one word, "no," but he said it in Swedish "Ney-da!", feeling so strongly about the issue he didn't think to translate.

"But you've just gotta listen to my reasons," Andrew blurted out. And without waiting for permission spoke this time without giving either parent an opportunity to interrupt, "I'm fifteen years old. I'm big for my age. I'm strong. You've said yourself, Pa, that I can do a man's work in the field. I've helped you cut down and saw up hundreds of trees. The logging company needs our horses. You really can get by without me this winter. Ma and Louise can handle the chores and before the winter's over you'll be back on your feet and able to work. And we really need the money to buy more land, now, it's like you said, Pa. If we wait another year someone else may buy part or all of that north 40 of Kuhlander's."

At this point, Andrew couldn't think of what he was supposed to say next, and he paused long enough for his parents to break in.

"Logging is just too dangerous for a boy," his mother said.

"And what about your education?" his father asked.

"Well, first off," Andrew replied, "I wouldn't be a logger, I'd be a cook's helper or take care of the horses."

Without giving his parents a chance for rebuttal, he went on, "And I don't know a single farm kid who has gone to high school. Eight grades is enough for the Olsons and the Kuhlanders and the Jensens and the Thompsons," Andrew added, listing the neighboring families who had children and then went on, "But this doesn't mean I won't go to high school; for now I'll just take a year off, I'm doing that anyway, you know."

Anders and Anna Anderson just looked at each other, so Andrew quickly added, "And you know the company is counting on Dan and Dolly (the family's team of plow horses), and we need the money we get for them just to make ends meet."

"Well," the father responded, "I've been planning to ask Nils Jensen to take them with him when he goes up."

Jensen was a farmer who lived on the other side of Upsala from the Andersons and who also worked for the Walker Lumber Company.

"But that wouldn't be fair to Mr. Jensen; he has his own horses," Andrew argued. "Besides, you'd have to give him part of the money for taking them."

"Ya, I know that," the father acknowledged. "But, Andrew, you are so young."

There was a sadness in his voice as he spoke, because other than the boy's age, his idea did make sense and could help avoid a setback in the family's plans and dreams.

Andrew detected a weakening in his father's opposition and quickly added, "I know I'm young, but I'm getting older and bigger everyday. I'll be sixteen by the time I get back home." Then looking at his father's protected, broken leg, asked with a grin, "Ya wanna wrestle? I'll bet I can beat you!"

Anders had to laugh at his son's phony challenge, but added, "Listen, you smart, young pup, when I get well I'll make you sorry you ever said that!"

Andrew sensed he was winning, so he turned to his mother and, knowing her concerns, assured her, "I'll be real careful, Ma."

But it was her turn. "It isn't just that I'm afraid you might get hurt, Andrew, but you might get sick. And then, too, I know a little about logging camps and know a lot more goes on than your father has told me, especially what happens when the men go into town, there's drinking, and fighting, and loose women."

"Mr. Jensen will take care of me," Andrew answered. "I promise to mind him just like he was my Pa. And I'll stay in camp."

"But is that fair to Mr. Jensen?" Mrs. Anderson asked.

"Well, we could *ask* him," Andrew suggested.

After nearly twenty years of marriage, Anna Anderson knew her husband pretty well and realized by the tone of his voice that he wasn't all that opposed to letting Andrew go to logging camp, so she signaled her reluctant permission by asking, "Well, Anders, what do you think?"

"Well," the stricken farmer replied, "first thing is we got to talk to Nils Jensen, and then if he agrees, we'll have to ask the foreman, John Schmitt in Little Falls, if he'll even *have* a job for Andrew."

The boy struggled to not show too much joy or enthusiasm. He could hardly believe what he was hearing; he had thought it would take days of arguing and pleading to get this far. "I'll go get Mr. Jensen first thing in the morning," he volunteered.

"No," the father replied, "we'll send Louise; I don't want you selling him on the idea against his will, the way you sold us."

"Speaking of Louise," the mother reminded her menfolk, "this will mean more work for her. We'd better see what she has to say."

"Let me talk to her," Andrew offered.

And Andrew did talk to Louise, and that was a mistake. He should have let his parents handle it. As a normal, typical teenage sister, she extracted promise after promise from her younger brother before she finally agreed, "Well, go ahead then, go to the woods, have your fun and adventure while I stay here and do chores, but don't you ever forget: you owe me!"

CHAPTER II

IT'S A GO!

Andrew didn't sleep very well that night when his folks gave him tentative permission to go to logging camp. Not only was he concerned that Mr. Jensen might not agree to his going along, but his mind was filled with thoughts about the logging camp and all the stories his father had shared over the years, stories about loggers, the heros of the north woods, about adventures, about snowstorms, even blizzards, about Indians, about Leech Lake, about food and fun in camp, about the boom town called Walker, and about Mr. Walker, himself, the founder and owner of the company for which he hoped to work.

When morning chores were finished, Andrew hitched "Flicka", Louise's horse, to the surrey. He even helped her into the four-wheel carriage as he pleaded with her, "Now please don't say anything to Mr. Jensen that might spoil my chances."

Actually, the request was totally unnecessary; Louise was mindful of how sweet it was having her brother obligated to her. As their eyes met she sighed, and then added, "Boy, do you ever owe me!" And then headed in the direction of Upsala.

Her instructions were to ask Jensen to come at his convenience, but when he learned of Anders' accident, he saddled up and returned with Louise. An hour or so later when he entered the Anderson bedroom, Jensen teased, "My goodness, what some people won't do to get out of work!"

As they shook hands, Anders thanked Nils for coming so quickly and assured him that he was on the mend. Andrew had been out by the barn when he had seen his sister and Mr. Jensen drive up and he had raced to the house. He didn't enter his parents' bedroom, but stationed himself in the hall where he could hear all that went on.

After exchanging pleasantries and making small talk, Jensen observed, "Well, Anders, I'm going to miss you this winter. I've always enjoyed our trips up and back and you've always been a good partner in camp."

"Same here," Anders replied, "but there are some things I won't miss, the lice, the aching muscles, the sore back, waking up with my hair frozen to the wall!"

Both men laughed, then Anders went on, "But as you may have guessed, I asked you to come because I need your help, a couple of favors."

"Well, let's hear them," came the reply.

"First off," Anders continued, "I know the company is counting on my team of horses. Will you take them up with you when you go?"

"Of course. No problem," Jensen answered. "What else?"

"Well, this is really a big one," Anders said hesitantly. "My boy, Andrew, wants to go in my place, not to be a logger; he's too young for that, but to be a "cookie" or to help with the horses. He really has it in his heart to go, and we do need the money. We want to buy one more 40 from the Kuhlanders before somebody else picks it up."

"I thought as much," Nils acknowledged. "Louise kind of let the cat out of the bag."

"Oh, no," Andrew thought to himself. "I wonder what she said. If she spoiled things, I'll...."

But Jensen's spoken words interrupted Andrew's momentary despair, "She said how much it meant to the boy and to you, of course I'll take him with me."

Andrew could have kissed her, almost!

"I'll go home by way of Little Falls and see the 'Old Bull of the Woods' (that's what the men called foreman Schmitt). I'm sure he'll have some work for your boy."

Andrew could restrain himself no longer. He burst into the room, grabbed Jensen's hand and said, "Thank you, thank you! You'll never be sorry, Mr. Jensen. I'll do everything and anything you tell me."

"Andrew," the father asked stearnly, "were you eavesdropping?"

Andrew's relatively dark complexion turned beet red; he was so embarrassed. He tried to talk, but couldn't.

The men could not help but laugh at the boy's predicament. "Well," the father went on, returning to his serious manner, "don't forget what you just promised Mr. Jensen. He has my permission to act as your father, even if that means giving you a good licking if you misbehave."

"Sure, Pa, I understand."

"I'm sure we'll get along just fine, Andrew," Jensen assured him. "I'll get back to you just as soon as I can, once I get the word from Mr. Schmitt."

At this point Mrs. Anderson brought in some warm apple pie and a pot of hot coffee. To Andrew, it was like a celebration.

Nils Jensen was as good as his word and was back at the Anderson farm the very next day with assurance that Andrew would be welcome in camp.

"You'll be a 'cookie'," He told Andrew. "You'll work hard, but you'll probably get fat sampling all that good food."

"Fatter," Louise snickered. It was true Andrew was a little on the chubby side.

What Nils didn't tell Andrew was that he had lied about the boy's age, or at least exaggerated by one year, and that the Bull of the Woods had extracted a promise that he would be totally responsible for the boy and for his behavior. He did tell Anders, however, "If it were not for Mr. Schmitt's high regard for you, I'm not so sure he would have agreed to take on your son."

Then, turning to Andrew he said, "So you see, you're going to have to live up to your father's reputation."

Andrew gave the expected assurances and then asked, "When do we leave?"

"I've given that a lot of thought, Andrew. How does one week from tomorrow sound?"

"Just great, Mr. Jensen. I could be ready tomorrow if I had to."

"Not so fast," Anders cautioned. "You've still got some chores to do to get this place ready for winter. You'll need every day of this next week, I promise you."

At first, Andrew thought a week was an impossibly long time to wait, but as his father had said, there was a lot of work to do: buck, chop, and stack more firewood, pile straw "cemented" with cow manure around the foundation of the house to keep out cold drafts, haul milk to the railroad station, take two more wagon-loads of corn to the elevator at

Upsala, take a load of eggs to Little Falls and bring back the things his mother needed from the store, and a whole lot more. By nightfall each day, Andrew was too tired to ask his father all of the questions that kept coming to mind while he worked during the day, but he did his best to stay awake during their evening visits. Anders thoroughly enjoyed answering questions about the logging camp and told story after story until the boy's head nodded and he could no longer keep his eyes open.

Andrew heard some pieces of advice from his father over and over, and these he promised not to forget, like:

"Always do what you are told and a little bit more."

"Be in the kitchen ready to go to work a few minutes before you're supposed to be there."

"Don't quit work for the day until everything you are expected to do gets done."

"Don't talk back to anyone; everyone there will be older than you."

"When you do outside chores, dress warm enough, but not so warm you will sweat."

"Do whatever Nils Jensen tells you, but don't be afraid to go to him with a problem."

"Some men, like "Black Richard", are mean. Stay out of their way."

"Stay out of fights, there's no way you're going to win. Those guys are tough."

To all of this, his mother added, "Wash your socks and underwear every Sunday, and when the minister comes to camp, be sure you go to the meetings."

When the big day of departure finally came, Andrew was more than ready. He had both horses tied out front. Dolly was saddled and Dan had all of the boy's needed possessions on his back, including a blanket roll, food to eat on the journey, extra clothing, his father's mackinaw (logging coat), a wash basin, soap, and a razor he'd bought in town just in case he started to grow whiskers over the winter. After all, when he looked in the mirror he could see a shadow forming on his upper lip.

When Nils Jensen arrived, the whole family saw them off, including Anders who, by now, was getting around fairly well on his homemade crutches. Andrew wasn't surprised that his mother cried as she hugged him good-bye, but he was shocked when Louise also gave him a hug and wiped away some tears.

After Anders shook hands, he pulled his old, beat-up logging cap, red plaid with a black tassel on top, from his hip pocket and said bruskly,

"Here, wear this for luck."

Andrew was too excited to feel bad or cry. His mind was totally on the adventures ahead. He was only fifteen, but as they rode off, he said to himself in all seriousness, "I think I've just become a man."

CHAPTER III
HEADING NORTH

"Where will we spend the night?" Andrew asked his older traveling companion.

"Hopefully, Brainerd." came the reply. "If we don't get there too late, I'd like to go into town to see my sister and her husband who live there, if you don't mind staying wherever we camp with the horses."

"Not at all," Andrew answered.

Conversation lagged as each buried himself in thought about what they were leaving behind and what adventures might lie ahead. Nils wondered how his wife would do with their three children, all school age but not yet in their teens. In recent years her younger brother had helped out each winter in Nils' absence, but now he was married and had plenty of responsibilities of his own. Nils' wife and the kids had insisted they could handle the chores, but it meant being up before daylight each day to feed the livestock and milk the cows and still get the children to school on time. They had to walk nearly two miles to the one-room schoolhouse. And then there would be chores after school but before supper. Nils felt guilty leaving them with so many responsibilities but, like his friend Anders, he knew it would mean a better life for his family in the long run if he could earn those extra dollars in logging camp. After all, he would be paid a dollar a day for his labors and another dollar each day for the horses. But he missed his family already. And what about logging camp? What if it was another winter of below zero weather and waist-deep snow, like last year? He could feel the aching muscles now. But there would be good times, too. He looked forward to seeing old friends, the story-telling in the bunkhouse, and even the satisfaction that comes from hard work well-done. It was important work. This was a growing country. Lumber was needed to build houses and schools and churches and factories.

Andrew, meanwhile, was doing his own daydreaming. He really didn't give much thought to what he was leaving behind. His mind was filled with expectations of what might lie ahead, his first real job for pay. Seventy-five cents a day wasn't anything to sneeze at. And he would be making new friends, living in a bunkhouse, and learning all about logging. But there were questions, too:

Would the men like him? Would there be anybody near his age? What kind of boss would the cook be?

A sudden change in the weather brought the travelers back to reality. They had left the Anderson farm with the sun shining and a gentle breeze from the south. The leaves were in full color, maple reds, popple and cottonwood yellows, and oak browns. As the trail led by occasional lakes and streams, they reflected the deep, rich blue of an October sky. But now the wind was in the north and the temperature was dropping. Clouds had turned the sky into a leaden gray, and leaves were being ripped from the trees and sent swirling across the trail in miniature whirlwinds. As Nils buttoned his shirt collar around his neck he looked back at his young companion, who had fallen behind a few paces, and noted, "I sure hope this doesn't mean an early winter."

And so, heads lowered into the wind, the riders continued their steady pace north, stopping only briefly around noon to eat a lunch of cold chicken and buttered rye bread and to stretch their muscles, stiff from immobility. Then they were on their way.

About mid-afternoon, Nils suggested another stop. "After all," he explained, "we seem to be ahead of schedule and the horses have earned a rest."

The spot Nils had chosen overlooked the joining of the Crow Wing and Mississippi Rivers. Actually, Nils had more than rest on his mind. He had chosen this place so that he could share some exciting Indian history with Andrew. "Are you interested in hearing about an Indian battle that took place here many years ago?"

"You bet," Andrew replied.

"Good, let's sit down on this log and have some of that cake my wife sent along and I'll tell you all about it. Do you see those pits across the river on the hillside?" Nils asked as he pointed out over the Mississippi.

"Ya," Andrew replied. The pits were easily visible to the naked eye.

"Well," Nils went on, "many years ago, the Sioux Indians controlled nearly all of Minnesota, but the Chippewa came here from back East with the French voyageurs and traders when they came to this area. The two

Indian tribes got along pretty good at first, the Chippewa helped the French trade for furs, but after awhile there was trouble and eventually all-out war. The Chippewa finally pushed the Sioux out of the timber areas of Minnesota and drove them south, but Sioux and Chippewa war parties kept raiding each other's villages for many years. And that brings me to the battle that took place right here."

"How do you know all this, Mr. Jensen?" Andrew interrupted.

"There's an Indian chief up on Leech Lake, Chief Flat Mouth, who told us this and a lot of other interesting stories during his visits to our logging camp. His father, who was also called Chief Flat Mouth, told him."

"Go on with your story," Andrew urged. "I won't interrupt again."

"Well, it seems there was this war party of Chippewa, about 70 braves, who came down the Mississippi River, right past here. They were headed for a Sioux village at the mouth of the Rum River where it flows into the Mississippi. When they got there, the village was deserted. Then an awful thought struck them; they got to thinking that the Sioux warriors might have gone north to raid their village and that was why the Sioux women and children had been hidden somewhere. They figured the Sioux might have taken the Crow Wing-Gull River route north and that was why they had not met up with them. Well, they guessed right. The Sioux had found the unprotected Chippewa village, up on Sandy Lake, and killed everybody except about 30 young women that they took as slaves and one older woman to help take care of them."

Nils paused to take a breath, and then went on, "So, anyway, the Chippewa headed north, expecting to meet the Sioux along the way, but when they got right here, where we're looking, and they still hadn't met them, they decided they'd better dig in and stay put, because they weren't sure whether the Sioux would be coming down the Crow Wing or the Mississippi. As it turned out, they didn't have long to wait. Early the very next morning, here came about 200 Sioux warriors and their captives right down the Mississippi."

Nils pointed for effect, paused as though waiting for Andrew to visualize the awesome fleet of canoes, and then continued, "The Chippewa were hidden in those pits, you see. But they had to wait, because the Sioux stopped on that island, right there, and made breakfast. Finally they got back in their canoes and the current brought them right in close to the bank there where the Chippewa lay waiting. Now as the story goes, the old Chippewa woman had whispered to the others that there was a good chance they'd meet the warriors from their own village somewhere along the way and told the girls that when they heard gunfire, to tip over the canoes and swim towards the shooting. Well, that's just what happened. When the Chippewa opened fire, the young women jumped up, turning the canoes over, and swam towards that shore where the Chippewa were."

"Wow!" Andrew exclaimed.

Nils went on, "The Chippewa had the advantage of surprise and even though they were badly outnumbered, about 200 to 70, they killed many. The Sioux that survived still outnumbered them and they tried to go around and attack from behind, but the Chippewa mowed them down before they could get to the pits. So the Sioux that were left finally gave up and headed down river."

"Was that the last battle between the Sioux and the Chippewa?" Andrew wanted to know.

"Oh, no," Nils replied. "In fact, there was another battle a few years later up the Crow Wing (Nils pointed to the smaller stream) where the Partridge River comes in. The Sioux attacked a French trading post and the Chippewa who wintered there. But that's another story, and we'd best be on our way and get across the river and head for Brainerd. I'd like to get to my sister's before dark."

As they mounted up and headed into the water, Andrew asked, "Do you think I'll have a chance to meet Chief Flat Mouth and hear more of his stories?"

"There's a good chance he and some of his men will show up in camp," Nils responded.

"Do the loggers get along well with the Indians?" Andrew wondered.

"Oh, yes, especially with Flat Mouth's village. But there have been some troubles between the government marshalls and the Indians from villages on the south end of the lake."

Little did Nils realize that even as they spoke, trouble between the Whites and the Indians was brewing, and three train-loads of soldiers were already at Walker. The last White-Indian war in the United States was about to break out!

A couple of hours later, the travelers made camp for the night on the east bank of the Mississippi on the edge of Brainerd.

"You can heat a jar of soup and make some sandwiches for your supper, Andrew," Nils suggested. "I'll eat at my sister's. It will probably be well after dark when I get back."

With that, Nils left Andrew with three of the horses and headed into town.

The chill in the fall air convinced Andrew he would need a fire to stay warm as well as to heat his supper, and it wasn't long before a cheerful fire was blazing and the kettle of soup was bubbly hot. Andrew finished his supper and was just washing up the cast iron pot and steel bowl and spoon when he became aware of a man on horseback heading his way in the twilight, from out of Brainerd. As the man drew near, he saw that it was not Nils but a very big man with a full, black beard. As he drew within speaking distance, Andrew said, "Hello."

The big man just glared at him and grunted. It was then that Andrew realized that the man was very drunk and having difficulty staying in the saddle. The man and horse proceeded up river about 50 yards, and then started to cross. Andrew watched, wondering if the rider would be able to stay on his horse as the river got deeper.

Suddenly, it happened. The horse stepped into a drop-off and the man rolled into the water. "Help!" he hollered, "I can't swim!"

Andrew knew better than to try to swim out to rescue a drowning man who was not only drunk but huge, so he carefully waded out, testing the bottom as he went. Fortunately, the hole into which the horse had floundered was not a long one and Andrew found himself standing in only waist-deep water as the current brought the man to him, his arms and legs threshing and throwing water in all directions. As he came by, Andrew grabbed the man's coat collar and said, "Relax, I've got you."

The big fellow stopped his struggling and let Andrew tow him to shore. Had he not been so drunk, he could easily have walked out under his own power shortly after he had drifted out of the hole.

The fear of drowning and the cold water had a sobering effect on the man and after he dragged himself to the fire, he raised himself up on one elbow, reached out to shake Andrew's hand and gasped, "Thank you, son, you saved my life."

Andrew didn't know what to say, but finally stammered, "Glad I could be of help, sir."

Meanwhile, the man's horse had returned so Andrew tied him next to the other three.

Moments later, the man asked, "What's your name, boy?"

"Andrew Anderson, sir."

"Well, I'm much obliged, Andrew. They call me *Black Richard*."

CHAPTER IV
WAR ON LEECH LAKE

Minutes after Andrew learned that the big man he had rescued was the infamous Black Richard who everyone in logging camp feared, the one his father had especially warned him about, Nils Jensen came riding out of the gloom. Nils recognized Black Richard immediately and couldn't help but notice he was soaking wet. As he dismounted, he observed, "Well, Andrew, I see we have company," and then turning to the fallen drunk said, "Hello, Richard."

Without returning the greeting, the big man simply said, "Nils, I owe this young man my life. I fell off my horse out in the river. I can't swim. He's a brave boy. Is he a friend of yours?"

"Yes," Nils replied, "he's Anders Anderson's son. Anders won't be up this year. He broke his leg."

"Sorry to hear that," and then looking at Andrew, he continued. "Your father's a good man."

The big fellow was shivering, even by the fire, so Nils directed, "Get out of those wet clothes, man; we'll wrap you in a dry blanket."

Black Richard nodded his agreement, but in his condition needed the help of both Nils and Andrew to get his clothes off. Andrew marveled at the size of the naked man and envied his bulging muscles. "No wonder people are afraid of him," Andrew thought to himself. "I sure wouldn't want him mad at me."

As Nils boiled some coffee, he asked Black Richard if he had heard about the trouble at Leech Lake. "Ya," came the reply.

"What trouble's that?" Andrew asked.

"Well, that's the reason I came back so soon and brought that rifle and shotgun with me." He pointed to a saddle holster on his horse, and then added, "It seems there's a war going on. The U.S. Marshalls had captured Chief Bug;* they wanted him to testify about some illegal sale of liquor to the Indians. They even had him in chains in Walker. He tried to run away, but they caught him, and that's when all the fighting broke out. The other Indians helped him escape. So, anyway, the marshalls were pretty badly outnumbered and sent for help. Three troop-trains of soldiers have gone up there."

Nils paused to pour some coffee all around and Andrew asked impatiently, "So what's happening?"

"Well," Nils replied, "it doesn't sound good. The soldiers took off on barges for Sugar Point, where old Bug is headquartered, and people could hear a lot of shooting, but so far the soldiers haven't come back to Walker. They don't know if they've all been killed or what."

"Golly!" Andrew commented, with eyes as big as saucers.

"The mayor of Walker telegraphed the mayor of Brainerd and asked for help. So a citizen's militia has been organized and they went up to Walker, also by train," Nils pointed at the railroad bridge a short way up river. "And no one's heard anything since. There's a lot of fear in Brainerd that the Indians might attack here, next."

"So what are we going to do?" Andrew asked with real concern.

"One thing's for sure, we're not going to head north until we find out what's going on up there."

At this point, Black Richard spoke up, "Well, I'll tell you what I'm going to do. I'm going to go to sleep." And he rolled over in his blanket and did just that.

"I don't think there's any danger here for us," Nils observed, "but that's why I borrowed the rifle and shotgun from my brother-in-law, just in case." With that he walked over to the horse and brought back both guns, handing the 12-gauge and a handful of shells to Andrew.

"So do we dare sleep?" Andrew worried.

"Oh, I think so," Nils answered. "You go ahead and I'll take first watch. I'll wake you when I'm tired."

For Andrew, it had been a long, strenuous, and exciting day. He thought he couldn't possibly go to sleep, especially thinking about the Indian uprising, but he dropped off in a matter of minutes. Nils meant well, but it had been a big day for him, too, and his watch lasted less than an hour as he also dozed off into a deep sleep without disturbing Andrew.

Andrew and Nils didn't wake until well after daybreak, when the noise

**The chief's full name was "Pugona-geshig", but the Chippewa "P" is pronounced "B"; hence, Chief Bug.*

of Black Richard making coffee aroused them. "Thought you two were going to sleep all day!" he grunted as the two stirred.

Nils joined the big man in making breakfast and Andrew asked question after question about what might happen next and what effect all this might have on the logging operations. There wasn't much either man could say, except, "We'll just have to wait and see."

About mid-morning the whistle and rumble of a train could be heard in the distance, heading for Brainerd from the northwest. Then they saw it, crossing on the trestle over the river. "It's a troop-train!" Nils exclaimed, "Let's get into town and see what's happened."

They broke camp in minutes and, with pack horses in tow, galloped for the depot. The train was already in the station when they got there, and a crowd of Brainerd citizens were quickly gathering. As they drew close, they could see a number of soldiers, who apparently had been wounded and were bandaged accordingly. In several cases the bandages were very bloody. Nils pointed grimly to the open doors of the baggage car where several ominous looking rough pine boxes were strapped in place, each, all too obviously, containing a dead soldier. A number of armed civilians were getting out of the railroad cars. A bystander identified them as members of the Brainerd militia. Two of the civilians, each carrying a rifle, got up on a platform and the crowd gathered around them. The bystander said, "The man on the left is Dr. Camp; the other man is Mayor Nevers."

The doctor and the mayor took turns speaking.

"All the Brainerd men are safe, but six soldiers, including Major Wilkinson, are in the boxes back in the baggage car."

"We have ten wounded soldiers," the doctor added.

"Is the war over?" someone yelled.

"We think so," the mayor responded.

"Tell us what happened," someone demanded.

Mayor Nevers turned to Dr. Camp and said, "Go ahead."

It turned out the doctor had been elected as head of the militia, so it was appropriate he give the report. He began, "When we got to Walker, we learned that the soldiers had gone to Sugar Point, looking for Chief Bug and his men. Earlier, the Walker people had heard a lot of shooting in that direction, but no one had come back. They were afraid all the soldiers had been killed; that's why they had sent for us. So we got all of the women and children into the Walker Hotel where they could be more easily defended, and we set up watchers on the trails leading into town.

And that's how we spent the first night."

No one interrupted him, and the doctor continued, "When nobody came back the next day, we took a barge and headed for Sugar Point. At the narrows, before entering the big lake, we saw a band of Indians on shore. We approached carefully, but they made signs of peace. It was Chief Flat Mouth and some of his braves and they asked where we were going. They seemed real concerned about the uprising and said they wanted no trouble and would be there when we returned.

"When we approached Sugar Point, we could see some men in uniform running down to meet us. At first we were afraid they might be Indians in disguise, but they turned out to be soldiers and they told us what had happened.

"When the soldiers had gotten to Sugar Point, there were no Indians in sight. There was a big clearing around a log cabin that belonged to Chief Bug. It was lunch time, so the soldiers were directed to stack their rifles in the center of the clearing and make camp. Just then one of the rifles fell down and went off. The soldiers didn't know it, but the woods were full of Indians and at least one of them thought they'd been discovered and opened fire. Then all the Indians began shooting. Like we told you, six soldiers, including the commandant, Major Wilkinson, were killed, and ten soldiers were wounded. As many as could got into the cabin and returned fire from there. After awhile the shooting stopped and the Indians went away. We really don't know if any of the Indians were shot or killed, but no one has seen them since.

"Anyway," the doctor concluded, "the rest of the soldiers are going to stay on for awhile to make sure the war is over."

As Dr. Camp was talking, Nils spotted several other loggers in the crowd who were also headed for Leech Lake. After the speech, he gathered them together to discuss what they should do. Dr. Camp and Mayor Nevers were also consulted. Knowing that the troops would remain on the lake for a time, it was decided that it would be safe to head north, but as a group, nine loggers in all, and each would carry a weapon.

A couple of hours later, as they rode out of town, Black Richard brought his horse along side of Andrew's. "Are you scared, kid?" he asked.

"Kind of," Andrew answered.

"Well don't you worry, son. I'll take good care of you. I owe you one, you know."

As the big man urged his horse ahead he looked back at Andrew and actually gave him a wink!

Andrew felt a whole lot better, not only because of the big man's words of assurance, but, as he said to himself, "Golly, I've got the meanest man in camp as my friend!"

CHAPTER V
LOGGING CAMP AT LAST

The lumberjacks headed out of Brainerd, crossed the Mississippi (this time they didn't fall in any holes!) and followed the trail north towards Walker, not stopping to rest until they hit the east end of Gull Lake about noon. As each man made his own lunch, Nils told Andrew, "There's a lot of Indian history right here, too. This is the old site of Chief Hole-in-the-Day's village."

"I'm not sure I really want to know, but did anything special happen here?" Andrew asked, hesitantly.

"Well, yes," Nils went on. "You must have heard about the big massacre of almost 500 Whites by the Sioux down around New Ulm during the Civil War?"

"Sure, we learned about it in school," Andrew answered.

"Well, at that very same time, the Chippewa up here had their own uprising, led by Hole-in-the-Day. They burned a mission church on this lake, not far from here, and captured all the Whites they could find between here and Leech Lake. Of course, in those days, that wasn't very many, but they brought them all here to this spot, and considered killing them. Some of the Whites got away without being captured and took refuge at Fort Ripley. You know where that is, between Brainerd and Little Falls?"

"Ya, I've been there," Andrew answered.

"Well, anyway, while they were debating what to do with their captives here, another Indian Chief by the name of 'Bad Boy' and his men from Mille Lacs Lake along with Father Pierz came here to confront Hole-in-the-Day."

"Why would Bad Boy side with the white people?" Andrew asked.

"As the story goes," Nils responded, "Bad Boy and his people used to live on Gull Lake, but they didn't get along very well with Hole-in-the-

Day, so it was kind of a chance to get revenge. Anyway, Bad Boy and Father Pierz talked Hole-in-the-Day into letting the white people go free."

"Wow, you sure know your Indian history, Mr. Jensen. Did you learn that from Chief Flat Mouth, too?" Andrew asked.

"Yep," was Nils' simple reply, "and now we'd better get going."

The lumberjacks made camp that night by Pine River, and Andrew slept with his borrowed shotgun by his side. He had heard too many Indian war stories for one day.

The next morning, after breakfast, the loggers mounted up and headed for Walker. The journey was pleasantly uneventful.

At Walker they found the troop-trains still in the station and the little town overrun with soldiers. Many other 'jacks' bound for the Walker logging operations were also in town, waiting until they felt sure it was safe to head into the woods. They reported that a number of loggers, including foreman Schmitt and the head cook, had gone up to logging camp more than a week earlier, and there was concern about their safety. More new men arrived the next day, and since no Indians had been seen for several days and feeling there was safety in numbers, it was decided they would all travel to camp together. It was determined they would go by way of the tiny village of Cass Lake before heading into the Sucker Bay camp; they thought it might be wise to check with folks there for any unfriendly activity.

About halfway they came around a bend in the trail and found themselves face to face with a small band of Indians on foot. Panic gave way to relief when one of the loggers yelled, "It's Flat Mouth!"

A brief visit with the chief assured the loggers that they had little to fear. Flat Mouth explained that Chief Bug's quarrel was with the government and that the loggers need expect no trouble from him. Flat Mouth added, "Chief Bug and his braves go to Boy River, not on Leech Lake."

Andrew was a little disappointed that the Indians were dressed mostly in white man's clothing, no war paint, no feathered head-dresses, but he was impressed that he had seen the famous Chief Flat Mouth.

Feeling much better, the lumberjacks went on into Cass Lake; here they found the villagers still in panic, gathered near a make-shift fort they had hastily erected. After relieving the villagers' fears, the crew traveled a few miles east and then south to the Sucker Bay camp.

Andrew was so excited he couldn't contain himself. "Logging camp at last!" he exclaimed right out loud.

It was a bigger layout than Andrew had expected. In addition to the three long bunkhouses, there were several smaller buildings, including a combination cook shanty and dining hall, an office, a supply building, a shelter for horses, a blacksmith shop and forge, a shanty where saws were sharpened, and a root cellar for storing food where it would not freeze.

Nils led the way to the barn where the horses were quickly tethered and fed. "All right," Nils said, "let's get settled in."

But at that moment, the Bull of the Woods himself, John Schmitt, and a handful of other men who had arrived in camp earlier, came out of the buildings to greet the newcomers. They were unaware of the battle and listened attentively as the stories were shared.

When the excitement subsided, foreman Schmitt, a big, raw-boned man who was about as broad as he was tall, approached Nils and Andrew and said, "This must be Anders' boy."

After a handshake and some small talk, he said, "Bring your shotgun and follow me, Andrew. I want you to meet your boss."

Schmitt led Andrew to a heavily bearded man on the edge of the gathering almost as big as Black Richard, at least around his middle. "Bruce," he said, "here's your new flunkie, Andrew Anderson."

"Well," the cook said as he shook hands and squeezed the muscle in Andrew's right arm, "I think he'll do; he seems big enough and strong enough to do a lot of lifting and carrying."

"Yes, sir," Andrew assured him, "I'll be a good worker."

"Just remember," The Bull of the Woods said, "Bruce is your boss, so do what he says and you'll get along just fine."

"You betcha!" Andrew assured.

"Bring your things and follow me to the cookshack," Bruce directed. "You'll be living there with me and the other cookies, that's what you helpers are called, you know." Andrew nodded and Olson went on, "We get up before everybody else to make breakfast. By staying in a different building, we don't wake them."

As they entered the cookshack, Bruce pointed to a big iron triangle hanging from a tree. "One of your jobs will be to ring this dinner bell when

I tell you. If I have time I'll blow a horn instead. We call it the "Gabriel", after the angel by that name, you know. I don't let anybody else blow that but me."

In the cookshack, Bruce pointed to some bunks in a corner and said, "Take your pick, then come up front to the kitchen. I only have one other cookie here so far, and it looks like we've got about 50 hungry 'jacks to feed for supper."

In moments Andrew reported to the kitchen and discovered the other cookie to be a delightful wiry, older man with a great sense of humor, named George Wilson. Andrew liked him immediately.

What followed next would be remembered by Andrew as a blur of activity, as he peeled many pounds of potatoes and carrots and put them on to boil while Bruce and George prepared huge beef roasts and gallons of gravy. The lumberjacks kept stopping in to consume hundreds of doughnuts the cook and George had made earlier in the day and drank what must have been a thousand cups of coffee and tea.

Besides helping with the cooking, Andrew was assigned the job of cutting and bringing in firewood and stoking the big kitchen stoves. He learned he would also make dozens of trips each day to the outside pump for water. Andrew was relieved to learn that there would be three other cookies, but that good news was balanced by the knowledge that more than another 50 lumberjacks would soon be in camp.

Supper was served "family style" with huge bowls of vegetables and big platters of meat. When the food was on the table, Andrew was told, "Ring the bell."

Before he could take more than a half-dozen swipes at the triangle with the steel rod clanger, lumberjacks began pouring out of their bunkhouses and into the dining hall. As Andrew watched them eat, he noticed that no one paid much attention to table manners, and that there was almost no conversation except, "pass this", "pass that", or "bring me so and so".

Bruce explained, "It's a rule in Mr. Walker's camps. He figures if there's no conversation there's less chance to argue or to fight about things. It seems to work. It's just like his rule about no drinking in camp. Saves a lot of violence."

"Do only Mr. Walker's camps have those rules?" Andrew asked.

"Not really," Bruce replied, "In fact, a lot of alcoholics go to work in logging camps to dry out."

As Nils Jensen left the dining hall, he sought out an obviously tired Andrew and asked with a twinkle in his eye, "Are you having fun, yet?"

After dinner the cooks had their chance to eat, and Andrew ate more than the other two put together, in spite of the fact he had "snitched" doughnuts all afternoon.

Then came the dish washing! What seemed like tons of steel plates and cups and all of the cooking utensils, which had been soaking while the cooks ate, had to be washed and dried.

Andrew was one very tired young man as he crawled into his bedroll on top of his hay-filled mattress on the bottom bunk. In seconds he was sound asleep, but his dreams were filled with thoughts of Chief Flat Mouth and Chief Bug and the soldiers and, of course, logging camp.

CHAPTER VI

A NEW BUDDY

Morning came all too early as big Bruce, the head cook, blared into Andrew's ear, "Daylight in the swamp, greenhorn!"

Andrew rolled out of bed, dressed in a daze, then stumbled into the kitchen, dimly lit with kerosene lamps, asking, "What shall I do first?"

"Fill those big coffee pots with water, put 'em on the stove, and when they come to a rolling boil, put six handfuls of coffee, it's over there, into each pot. They'll stop boiling for a few minutes; when they start to boil again, pull them over to a cooler spot on the stove so they won't boil anymore but will stay hot. Then you can bring in some more firewood and stoke the stoves."

"What's for breakfast?" Andrew asked as he worked.

"Stove lids and sausage", came the reply.

"Stove lids?" Andrew asked.

"You'd call them *pancakes*," George Wilson explained. "We also call them *flapjacks*."

By the time Andrew had the coffee cooked and had hauled in more firewood, Bruce and George had stacks of cakes and platters of sausage ready.

"You can ring the bell, Andrew, and then start throwing food on the table," the cook ordered.

Andrew hadn't finished clanging the triangle when the stampede of hungry men started pouring into the dining hall.

After breakfast, the lumberjacks were organized into crews to get equipment ready for the logging operation which was scheduled to start in a few days. After finishing breakfast dishes and re-setting the tables, Andrew helped the cook stamp out doughnuts. Meanwhile, he began asking questions about logging camp vocabulary. "Calling pancakes stove lids kind of threw me," he told Bruce. "What are some other new words I should know?"

Let's see if George and I can help you," Bruce replied, and the two of them came up with quite a few sayings like:

"Coffee is usually called *java* or *black jack*. Strong coffee is *lye*."

"Milk, when we have it, might be called *cow*."

"Eggs are often called *cackle-berries* or *hens' fruit*."

"If someone wants *gravel*, that means they want salt."

"Pepper may be called *Mexican gravel*."

"Sugar is *sand*."

"They have a lot of different names for beans, like *whistle-berries* or *fire crackers*."

"*Mulligan* means stew."

"If someone asks for *Adam's fruit* or *Eve's fruit*, that means apples."

"The general word for food is *grub*."

"Potatoes are called *spuds*."

"*Sow belly* means salt pork and *red horse* means roast pork."

Bruce concluded, "I'm sure we've missed some, but you'll pick them up by and by."

About mid-morning, new men started arriving, most with horses but some with teams of oxen.

Shortly before lunch, the back kitchen door was flung open, and a lanky, good-looking youth with curly black hair entered the room proclaiming, "You have nothing to fear; Tom Johnson is here!"

"Hey, Tommy!" the cook shouted back a welcome. "Glad you finally showed up; we can sure use the help!"

George Wilson hurried over and shook the newcomer's hand; it was apparent to Andrew that the newly-arrived cookie was well-liked, and Andrew was pleased to see someone nearer his own age. About that time, the young man spotted Andrew and greeted him with, "Hey, who've we got here? Someone younger than me, someone I can boss around for a change!"

"You won't be too tough on me, will you?" Andrew asked, extending his hand. "My name's Andrew Anderson, Tom."

"No more time to talk, guys," Bruce interrupted. "Help us make lunch, Tom; you can get settled in afterwards."

Everyone pitched in preparing stew, or mulligan as George Wilson called it, and sandwiches.

After cleaning up the dishes from the noon meal, Andrew offered to help Tom move in, anxious to solidify their new friendship.

"Sure," he responded, "I can use the help and we can get better acquainted."

As the boys brought the duffle in and headed towards the bunks in the corner against the wall, Tom suddenly stopped short and asked, "Hey, who stole my bunk?"

"Which one?" Andrew asked.

"That one," Tom said, pointing at Andrew's bunk.

"Sorry," Andrew said, "I'll move to the top one."

"You were here first," Tom admitted. "Tell you what; I'll wrestle you for it!"

Andrew studied his new friend. There was no doubt he was a couple years older, but he wasn't quite as tall or as heavy. "You're on!" Andrew yelled, and tackled the older boy around the waist. They went crashing to the floor, Andrew on top. Andrew stayed on top for awhile, resisting Tom's every move to roll him over and using his weight to good advantage. When he sensed that he was wearing Tom down, he decided to finish him off by sitting on his chest and pinning his wrists to the floor, but that proved to be a fatal mistake. Andrew learned the hard way how quick and strong Tom was, as the older boy suddenly pushed up on both of Andrew's knees, sending him sprawling on his back. Tom was on him like a cat, wrapping his legs around the muscle of Andrew's right arm, laying across his chest, and then easily controlling Andrew's left arm with both of his hands. Tom pushed down with all his weight, making it hard for Andrew to breathe. He bucked and twisted but the more he squirmed, the tighter Tom cinched up his hold. Finally, realizing there was no way he was going to get Tom off, he wheezed, "I give."

"Not bad, Andrew," Tom said as he extended his hand and pulled him to his feet. "How old are you anyway?"

"Sixteen," Andrew lied.

"What do you weigh?"

"Just before I left I weighed 165 on the scale at the grain elevator."

"Well," Tom said, "you've got me by about 10 pounds, but I just turned eighteen, so you did well."

"If you're eighteen, you've probably stopped growing. You'd better watch out. I'm bigger than you now; another inch or two and another 10 pounds and you've had it!" Andrew bragged.

"Then I'd better make hay while the sun shines," Tom yelped as he threw his shoulder into Andrew's soft mid-section and pinned him on the lower bunk, holding fast to the bedframe under the mattress, pulling himself down hard on his younger opponent.

Andrew didn't have a prayer and he knew it, so he held out a respectable time and then admitted, "You win again, Tom."

Thus it was that with a couple of good-natured wrestling matches, a new friendship was cemented.

Since it was a little too early to start supper, Tom gave Andrew a tour of the logging camp, introducing him to friends along the way. Andrew noticed that both walls of the bunkhouses were lined with double bunks and there was a pot-bellied stove in each end of each building. One bunkhouse was reserved for the teamsters, those who handled the horses and oxen. "Any special reason?" Andrew asked.

"Sure," Tom replied, "in a couple of weeks, the men and this place will smell just like the animals!"

Andrew found the blacksmith shop especially interesting, and the boys watched the smith and his helper forge an iron chain.

As Tom introduced him to logger after logger, Andrew became aware of a variety of accents; English, Scandinavian, German, and others he couldn't identify. Tom pointed out that many had what he called "Maine accents". These were professional loggers who had begun their work back east in New England, and worked their way west as the trees were harvested. About mid-afternoon, Tom observed, "We'd best be getting back to the cook shack or Bruce will be after us."

The next day, two more cookies arrived, both in their late twenties, and the crew was complete.

A few days later the camp awoke to a cold north wind and snow in the air. That night the wind went down but the temperature plummeted to well below zero. The following morning Bruce observed, "It really made ice last night; it should be safe for ice fishing. Maybe you boys would like to try your hand at getting enough pike for a meal."

"Sure," Tom agreed, "come on, Andrew, this will be fun!"

"How do we do it?" Andrew asked. "I've never fished for anything bigger than sunfish."

"I'll show you," Tom replied. "We begin by dressing plenty warm."

As the boys put on their mackinaws and leather boots, Tom explained, "We'll take clamp skates with us so we can get from hole to hole fast. Do you skate?"

"You bet. What do we use for bait?"

"Perch. We'll catch them first by using the little white grub worms found in goldenrod weed balls."

Pulling a sled with a box full of lines with hooks, two axes, a bucket, a scoop, and their skates, the boys took off down the trail that led to Sucker Bay on Leech Lake. Along the way they picked a couple dozen goldenrod bulbs, each containing a worm they would harvest later.

When they reached the beautiful bay, Andrew asked, "How come the logging camp isn't located on the lake?"

"The camp would be a lot colder," Tom explained. "It would be exposed to the winter winds. Then, too, our location back in the woods is more central to the logging operations."

Looking across the bay toward Ottertail Point, Tom asked, "Did you ever hear of Chief Flat Mouth?"

"Yes, in fact we met him and some of his men on the trail between Walker and Cass Lake. Why?"

Tom pointed towards Ottertail Point, "Do you see the north narrows where the treeline dips?"

Andrew said that he did.

"Well, that's where Flat Mouth's village is located. But so much for your geography lesson," Tom added. "Let's go fishing!"

Tom produced two lines, each with a small hook, tiny lead sinker, and wooden bobber. "Now," he said, "use your axe and cut two round holes about a foot across."

The four inches of ice proved easy work. Tom had chosen a spot where the water depth was about seven or eight feet. He set each bobber at six feet and showed Andrew how to cut open a goldenrod bulb, find the worm, and run it onto the hook. Seconds later, Andrew had his first bite, but he set the hook too soon, and missed. "Give it a little more time," Tom suggested, "Like this!" he yelped as he set the hook and in moments had a fat, six inch perch flopping on the ice.

Tom quickly scooped the bucket half-full of water and put his perch in it to keep it alive.

Now it was Andrew's turn, "Hey, this is fun!" he said as he tossed a perch a little smaller than Tom's into the bucket.

A half-hour and twenty or so perch later, Tom said, "That's enough. Let's go after the northerns, but first we have to go back to shore. I forgot to cut the willow sticks. We'll need ten."

After cutting the sticks, each about 18 inches long, Tom led the way farther out onto the lake, testing the ice from time to time to be sure it was safe. He stopped where the water was about ten to twelve feet deep, and announced, "We'll cut ten holes about 100 feet apart."

When this task was finished, he said, "Now, watch what I do."

He scooped the ice chunks out of the nearest hole into a pile. He then pushed the larger end of a willow stick into the pile of ice chips and poured a little water on the pile so the stick would freeze in. The tip of the willow was positioned over the center of the hole. Next he took some red cloth from the box, "Bruce's old underwear," Tom explained, and cut off a three-inch, narrow strip which he tied to the tip of the willow over the hole. Next he produced one of the sticks around which the heavy fishing line was wrapped. It had a big hook and a heavy sinker, but no bobber. He measured about eight feet of line, hooked one of the perch under its back fin, dropped it in the hole, and gave the line several wraps around the willow stick next to the red flag. He then laid the stick that the line was tied to across the hole where it could not be pulled in by a fish, should the willow stick be pulled out of the frozen slush.

"When a fish hits the perch," Tom explained, "the flag will be pulled down into the hole. So when we can't see a flag over a hole, we'll know we have a bite."

The boys continued to set lines in the other holes, meanwhile keeping an anxious eye on the flags that had already been set. Suddenly, Andrew noticed that the flag over the third hole was missing. "We've got a bite!" he hollered.

The fishermen had not yet put on their skates, so they took off on the run, each falling once before reaching the hole and both sliding past when they did reach it. Scrambling back, they could see the flag down in the hole, jumping around violently.

"He's still on, Andrew, grab the line and set the hook, hard."

Andrew did just that and yelped, "He's on!"

"If he pulls hard, give him line," Tom coached.

Andrew did have to play the fish for awhile, but in a few minutes he could see him below the hole. "He's a big one!" Andrew said excitedly.

"Naw, only about five pounds," Tom observed.

"That's big for me," Andrew argued.

Moments later the northern was flopping out on top of the ice and Andrew had his first trophy.

As they reset the line, Tom suggested, "We'd better get our skates on." So they clamped them to the soles of their boots before setting up the remaining holes.

From then on, Andrew had one of the most fun-filled afternoons of his young life. Whenever a flag disappeared the boys would race to the spot.

As it turned out, the five pounder was one of the smaller fish of the day. The biggest and last caught weighed in at 22# back at camp, and Andrew landed it after a long battle. The fish had raced back and forth below the hole and ran to the end of the line many times. Andrew was sure the line would break, but it held each time. At long last he pulled the fish's head through the hole. Tom took a firm grip behind the gills and slid him out onto the ice. Andrew was estatic!

Shortly thereafter, Tom noticed that the sun was only about an hour above the horizon, so he reluctantly advised, "We'd better head back. Bruce will need us to help with supper."

The box on the sled barely held the twenty northerns, but it was a happy load the boys pulled back to camp.

As soon as Andrew's biggest fish was weighed in, he tracked down Nils to show him his trophy. "Hey, Mr. Jensen, look at my big northern!" he shouted.

"That is a beauty, but it isn't a northern," Nils responded.

"If it isn't a northern, then what is it?" Andrew asked.

"Notice the markings? They run up and down instead of lengthwise like a northern. You've caught a muskie; in fact, it's a tiger muskie!"

"Is that as good as a northern?" Andrew asked innocently.

"It's better, Andrew. Muskies are the fightingest fish of all. It's a real trophy."

Nils had made a happy boy even happier. When the muskie was butchered, Andrew nailed the head to a tree where he could look at it everytime he went out for water or firewood.

CHAPTER VII

BLACK RICHARD TO THE RESCUE

Within a few days after the fishing expedition, the logging operations were in full swing. With well over a hundred hungry lumberjacks in camp, the kitchen crew had their hands full. Andrew was used to hard work on the farm, but this was work of another kind. It called for muscles he didn't know he had, and by the time the last pot was washed each night and the tin dishes were on the table ready for the next day's breakfast, Andrew was sore all over, bone tired, and ready for his bunk. He couldn't recall ever sleeping so well, and the cook's wake-up call always came as a shock. The nights seemed so short.

Because this was the third winter the logging camp had been in this particular location, the crew had to work more than a mile from camp. Another year and it would be time to tow the smaller buildings to a new site. The bunkhouses were too big to move, but the doors, windows, and furnishings would be removed and hauled to the new location. Since the crews were working too far from camp for the men to come in for lunch, it was Andrew's and Tom's job to haul the food out to the lumberjacks on a specially outfitted sled called a "swing dingle", drawn by a single horse. Lunch consisted of soup or stew, which usually wasn't very hot by the time the last logger got his share, and huge sandwiches made of thick slices of sourdough bread with slabs of beef, pork, or ham. And of course, there were the ever-present doughnuts for dessert.

Andrew was thrilled to watch the loggers at work. He had helped his father cut down many a tree, but they were mainly oak and cottonwood. Here in the northern forest they were cutting towering white and Norway pine, most so big he couldn't reach his arms around their trunks. The lumberjacks were true professionals, and could drop a tree within inches of where they wanted it to fall. This was important so that they wouldn't hang up on other trees and cause a lot more work.

Because the two-man crews worked close together, it was important for their safety that everyone know when a tree was about to fall, and Andrew thrilled to hear each lusty call of "T - I - M - M - M - B - E - R - R - R !".

When the first Sunday after the first full week of work rolled around, everyone was ready for a break. Muscles had not yet been hardened, and the men were pretty well worn out and a bit owly and ornery. Sunday was the only day of the week when the big meal was served at noon, and that was the occasion of Andrew's first real problem in camp. Although he had been working very hard, he enjoyed his work and had gotten along well with the men. He especially liked working for Bruce, the head cook, and his new friendship with Tom had made the whole adventure even more enjoyable. But, sooner or later, something was bound to go wrong, and it was while serving dinner that Sunday that he had his accident. Andrew was carrying a huge bowl of steaming hot soup to the table when he slipped on a stretch of wet floor, made icy by snow dragged in on the boots of some logger. As he went down, the soup went up in the air and part of it came down in the lap of a muscular lumberjack known as Boomer Smith. He was called "Boomer" because he had moved so often from camp to camp. Rumor had it that he was usually fired because of his temper or because he couldn't get along with the boss or some of his fellow workers. This was his first year in the Walker camp.

As the hot soup hit Boomer's lap, he cried out in pain and jumped to his feet. Andrew got up off the floor where he had fallen and was starting to say how sorry he was when Boomer hit him across the face with the back of his hand, sending him sprawling back to the floor and up against the wall.

"Get up you clumsy brat!" he yelled at Andrew, with both his fists clenched and waving menacingly.

Before Andrew could decide whether he should obey or whether it might be smarter to stay down, who should come between them but Black Richard.

"Leave the boy alone!" he ordered. "Can't you see it wasn't his fault? He slipped on the snow."

"Listen, mister," Boomer Smith replied, "I don't know who you are, but if you want to make this your fight that's just fine with me."

Black Richard pulled himself to his full height, looked down his nose into Smith's face and roared, "Are you sure you want to mess with Black Richard?"

Smith didn't move back an inch, but answered, "I've done a lot of boxing. I'm county champ where I come from and I've handled bigger men than you. Are *you* sure *you* want to mess with *me*?" At this point, foreman Schmitt's voice boomed, "No fighting in the dining hall; settle your differences outside!"

"Fine with me," Black Richard answered, still glowering at Smith.

"See you outside after we eat," Boomer snorted.

Andrew, who was the cause of all the trouble, was forgotten; but he trembled with excitement and concern as he helped serve the rest of the meal.

"I'm impressed," Tom whispered as the boys passed each other as they worked. "I've never seen Black Richard stick up for anybody. How come he likes you?"

"Oh, I just saved his life once, that's all," Andrew whispered back as casually as he could.

Tom just stopped, scratched his head, and stared at Andrew in dumb amazement.

With dinner finished, the men gathered outside, anticipating the big fight. Most of them couldn't have cared less who might win. Few of them liked Black Richard, but Boomer's behavior in the dining hall had gone beyond all reason. Andrew asked Bruce if dishes could wait while they watched. The big cook replied, "No problem. I wouldn't miss this fight for the world!"

It seemed like every man in camp was there. Andrew and Tom had to crawl up on the roof of a bunkhouse to see the action, and they didn't have long to wait. Boomer showed first; when Black Richard appeared, the newcomer snorted, "OK big boy, let's have at it. I'm going to enjoy teaching you to mind your own business."

Black Richard just grunted.

In spite of below freezing temperatures, both men stripped to the waist. Andrew sized them up. He had seen Black Richard naked, so he wasn't surprised at his barrel chest, ample waist, hairy body, and bulging arm and neck muscles. Boomer was nearly as tall as Black Richard but probably weighed 50 pounds less. Yet he was muscular. His biceps were as big as his adversary's and his back muscles rippled when he flexed his arms. "Not an ounce of fat," Andrew whispered to Tom.

As the two men began to circle each other, Boomer moved with the agility and coordination of a trained boxer, dancing around his bigger opponent. Black Richard, in contrast, circled slowly and deliberately, flat footed, more like a wrestler.

Boomer struck first as he lashed out with a right to Black Richard's jaw, snapping his head back. Seconds later, he swung again, a left to his opponent's huge mid-section and a right to the mouth. A trickle of blood began to moisten Black Richard's beard.

The bigger man had not laid a hand on his smaller opponent, and it looked for the moment as though Boomer would cut Black Richard to shreds and make good on his threat.

But the next time Boomer moved in, Black Richard was ready. He ignored a blow to his stomach, met Smith's charge, wrapped his big, hairy arms around him in a bear hug, lifted him off the ground and held him there. Boomer's fists could no longer find their mark. The best he could do was beat his opponent on his shoulders and back. Black Richard just squeezed harder and harder until Boomer could no longer breathe. After what seemed like minutes, Smith stopped struggling, and Black Richard let the smaller man's feet touch the ground and then quickly bent him over backwards, held him there for a few seconds, and then dropped him to the ground with his own full weight on top, squashing more air out of his lungs.

Black Richard raised himself up on his knees, straddling Boomer's chest for control, and then powerfully and deliberately pummeled the man's face with his huge fists until he was unconscious. The victor slowly rose to his feet and stood for awhile over the vanquished, making sure he wasn't faking. Finally, satisfied that he had beaten him, he planted his huge hobnailed boot on the fallen man's face and pushed down with all his weight. He then picked up his shirt and walked off into the silent crowd. Black Richard was still "champ of the camp".

Andrew looked at Tom and said just one word, "Awesome!"

Tom nodded.

A little later, Andrew sought out Black Richard in his bunkhouse and thanked him for coming to the rescue.

"Well, lad, I owe you that and more, but try to stay out of trouble, will ya?" Black Richard requested.

"Don't worry, I will," Andrew assured him. "but I don't think Boomer Smith will bother me anymore."

"Do you know why I stamped on his face after I licked him?" Black

Richard asked Andrew.

"No," the boy admitted.

"Well, I know it looked a bit barbaric, but I just wanted to make sure every time he looked in a mirror he'd remember how bad I beat him. I don't want him ever to challenge me again or try for revenge."

"Oh," Andrew responded quietly.

"You may see a few other men around here with marks like measles on their faces," Black Richard went on. "Most of 'em got marked by my boots. But I'm not the only one that does it. You'll see that in nearly every logging camp. Some call it 'logger's smallpox'." Big Richard chuckled.

Sunday was laundry day, and Andrew joined most of the men in washing his long johns and hanging them outside to freeze-dry. Later some of the men took up games for entertainment. Some played poker, using dry beans for chips, or other card games. Mr. Walker had a rule against gambling, but a lot of money did change hands "under the table". Still other men played games like "Jack in the dark", "rooster fight", "squirrel", and "hot buns".

Tom introduced Andrew to the game called "rooster fight". Each boy squatted down holding a broom handle behind his knees, with both hands. The object of the game was to tip your opponent over so he would fall on the floor, but if a player let go of the broom handle with either hand at any time, he was declared the loser. Tom had played the game many times before and had a distinct advantage. By actual count, he sent Andrew sprawling ten times before the younger boy learned to balance himself and catch Tom with his shoulder.

Just when Andrew was getting the hang of it, one of a group of loggers who were dunking doughnuts at the table called to Andrew, "Want to learn a real logger's game?"

"Ya sure," Andrew replied.

Harry Henderson, the logger who asked the question, explained the rules: "First, we blindfold you. Then you get down on your hands and knees. We'll form a circle around you and we'll keep moving, taking turns swatting you across your rear end. You have to guess who's hitting you.

When you guess the right man, he has to take your place."

Andrew submitted to being blindfolded with his own big, red, logger's handkerchief, and then assumed the required position on his hands and knees.

He could hear the loggers shuffling in a circle around him and every few seconds he felt the sting of a hard swat on his rear end. Vainly he guessed, "Tom", "Harry", "Joe", "Arnie", "Luther", "Rueben", "Mike", but at each guess, everyone shouted, "No!", and laughed hysterically.

After many minutes and much pain, the horrible thought crossed Andrew's mind that they might be lying and would never admit he guessed who had hit him. He could hear Tom laughing uncontrollably and knew something was up, so he slipped off his blindfold, stood up, shook his finger at Harry, and said, "I don't think you'd tell me if I did guess right!"

The men doubled with laughter and Andrew knew he had guessed the awful truth.

"You're right, Andrew", Harry admitted, "but it is a real game and we do follow the rules, except when there's a greenhorn in camp."

Andrew managed a smile and asked, "What's the name of the game?"

"Hot buns!" Harry replied with a chuckle.

Andrew was reminded how appropriate the name was every time he tried to sit down the next couple of days.

"There's another bunkhouse game we play blindfolded," Harry told Andrew. "This time you can watch so you know it's on the up and up. It's called 'Jack in the dark.'"

Two men volunteered to be "it". They were blindfolded and got down on their knees. Each was handed a sock with a rolled-up, wet dish towel in the toe for weight. Luther was labeled "Jack in the dark". Rueben had to find him by asking, "Jack in the dark, where are you?" And Luther had to answer, "Here." Then both men started swinging at the sound they had heard. The swatting continued until Luther finally gave up, more from exhaustion than from any real beating.

Next, Tom and Andrew were "it". Andrew again learned to respect Tom's experience as the older boy would answer "here" in a small, squeaky voice, move noiselessly out of the way, and then score often and well as he swung at his noisier opponent. It was Andrew who finally gave up.

Later, when they were alone, Tom showed Andrew another game two can play, called "squirrel". Tom directed Andrew to take a seat on a bench. "Now place your hands on your knees and then spread your knees apart," Tom told him.

When Andrew was in position, Tom put his cap on his head and then knelt down facing him and explained, "I will try to pass my head between your knees while making a noise like a squirrel. Your job is to knock my hat off. If I succeed, and you miss, then I'll make the same noise and bring my head up. When you knock my hat off, we'll change places. The one who makes the most bobs without losing his hat wins. But remember, you can move your head only while making the squirrel noise."

Eight bobs later, Andrew finally knocked Tom's hat off. When they changed places, Tom again showed his experience by knocking Andrew's hat off the first try!

Andrew felt something like a winner, however, because in trying to knock Tom's hat off, he inflicted many a blow to the older boy's cheeks and ears!

Just before supper, Frank Higgins, a traveling minister or "sky pilot" as the loggers called him, arrived in camp. He was well known to the men and well liked. Although some of the loggers didn't especially enjoy his "hell, fire, and brimstone" sermons, they did enjoy the hymn sings, led by the pastor in his strong baritone voice.

After supper, most of the lumberjacks gathered in the cookshack for the service. Harry Henderson brought his violin, George Jones and Jerry Howe brought their harmonicas, and Bruce, the cook, took out his ukelele. Pastor Higgins gave the men a cheery word of welcome, opened with prayer, passed out the songbooks, and then led the men in an hour of such favorites as "Amazing Grace", "Power in the Blood", and "Battle Hymn of the Republic". Andrew loved to sing, and he knew most of the hymns by heart from attending services in the little Swedish Covenant Church back in Upsala. He sang along at the top of his voice, which was really quite good.

After the hour of singing and another hour of preaching, Andrew's head began to nod. It had been a long day, from Black Richard's beating of Boomer Smith to an afternoon of laundry and playing games, besides all his work as "cookie". Andrew was relieved when Pastor Higgins said his final “amen” and ended the evening by offering to “pray with any ‘jack who wanted to be saved”.

CHAPTER VIII

THE BLIZZARD

Sometime around midnight, Andrew was awakened by the howling wind. The bunkhouse was not insulated in any way and icy drafts slipped between the logs and around the windows, making the long, narrow building even colder than usual. The fires in the cook stove and heating stove had only lasted a couple of hours after the men went to bed. Andrew pulled his wool stocking cap down over his ears, buried himself a little deeper in his bedroll, and was soon once again in a deep sleep.

The cook's wake-up call came well before daylight, right on schedule. Andrew often wondered how Bruce awoke the same time every morning without the help of an alarm clock. When asked, his reply was, "I have a built-in alarm clock." Andrew was not sure what that meant. He only wished that just once in awhile the big man would oversleep.

As Andrew performed his first morning chores of going out for water to make coffee, he found he had to put his shoulder to the door to force it open against a big drift of snow that had formed overnight. The wind was still blowing fiercely and snowflakes swirled into the room through the open door. Luckily, the well was just a few steps from the building and Andrew was grateful he could turn his back to the wind as he pumped. Bringing in the first buckets of water he announced to the kitchen crew, "It's a blizzard!"

"Ya, I was afraid of that," Bruce acknowledged without looking up from the huge skillet in which he was scrambling eggs. "Could be the men will have the day off."

Foreman Schmitt was the first man in the cookshack that morning; he too had been awakened by the storm.

"What do you think, Mr. Schmitt?" Bruce asked. "Will the men be able to work today?"

"I doubt it," the Bull of the Woods replied, "but we'll see what daylight brings."

Snow or no snow, the hungry 'jacks poured into the dining hall on schedule, ready for breakfast. Eager for a day off, the men within earshot of the foreman cleverly remarked to each other as they sat down, in tones just loud enough so he would be sure to hear: "Pretty rough out there", "Must be a blizzard", and "This could be a big one!"

Daylight comes late in the Minnesota woods in December, but when it did come, it was plain there would be no work that day. Andrew peered out the window at first light and remarked to Tom, "I can't even see the woodpile or the blacksmith shop."

So the men had their day off. For them it was a time for playing cards and other games, telling stories, coffee and tea drinking, dunking doughnuts, and afternoon naps. But for Andrew and the cooking crew, there was no rest. The men seemed to eat even more than when they worked. The cooks particularly had a tough time keeping up with enough cookies and doughnuts for between meal snacks.

The wind was still howling out of the northwest as darkness came, but when Andrew awakened sometime in the night, he was aware that the wind had stopped blowing and he knew the storm was over.

As so often happens after a winter storm, the temperature dropped. That next morning the camp thermometer just outside the cooking shack door registered 30° below zero, but without the wind it would not seem that cold in the woods. As the men would dig out from the storm, there would be more sweating than freezing. Although it would be a day of work, few trees would be cut. Nearly two feet of new snow had to be shoveled away from the buildings; trails had to be cleared to the cutting areas more than a mile from camp; and the road to the lake had to be re-iced so that the horses and oxen could pull the heavy loads of logs. Once the trails had been cleared, specially designed sleds with tanks dribbled lake water behind them to make the road a solid sheet of ice. It was hard for Andrew to believe that the animals could pull such heavy loads, but the iron strips on the bottoms of the sled runners slid easily, there was so little friction. Of course, the animals were shod with cleated iron shoes so they wouldn't fall, and the ice was usually relatively rough. Then, too, the roads to the lake were laid out to avoid hills and to be as level as possible. In those days, animal-drawn sleds were the only way the logs could be brought to the lake where they could be floated to Walker after ice-out in the spring. A few years later, steam-driven crawler tractors and even locomotives running on tracks would replace many of the animals. Logs could then be hauled directly to the mills instead of being floated across the lake.

By the second day after the blizzard, logging operations were back to normal.

That next Sunday, after breakfast, Bruce asked Andrew and Tom, "Would you guys like to try your hand at deer hunting?" Assuming a positive reply, the cook didn't even pause for breath but continued, "I'm afraid the supply sleds from Walker are going to be several days late because of the storm and we're getting low on meat. The men aren't going to go hungry, but they could be a little hard to live with if we have to start rationing pork and beef until the supplies get here."

As he had expected, Andrew and Tom were more than willing to get a break from the camp routine, and the latter asked, "When can we go?"

"As soon as you finish breakfast dishes. You'll probably need all day. I'll get some volunteers to help with the noon meal; so do the dishes, pack some sandwiches, and take off."

"What do we do for guns?" Andrew asked. "All I have is a shotgun we borrowed in Brainerd from Nils' brother-in-law."

"Make some bows and arrows!" was Bruce's smart reply, but then he added, "There are a couple of 30-30's under my bed; the shells are there too."

So it was that after breakfast dishes had been washed and put back on the table, the young hunters headed for the woods, stopping briefly at the supply room for snowshoes.

"Any idea where to hunt?" Andrew asked.

"I think our best chance is to head for the big cedar swamp and follow the edge towards the lake. I know that deer eat cedar needles when the snow gets too deep to paw for food."

"What do we do, just walk until we see one?"

"With this deep snow the deer won't move around much. I think the best plan is to look for tracks and then each follow a deer. If we're real quiet and work into the wind so they can't smell us, we should get some shooting," Tom answered knowingly.

Even with the snowshoes the going was difficult. The new snow was so powdery it puffed up through the webbing, but it was far better going with than without them. The two boys walked the two miles west to the swamp without seeing a single track.

"I sure hope you're right about the deer being along the swamp, Tom," Andrew whispered. "I can't believe we haven't seen a track."

"It was what I expected," Tom replied, showing no concern.

But when they had followed the edge of the swamp for a mile, then two miles, then three, Tom stopped and admitted, "I don't know what to think. They should be here. We sure saw plenty of tracks and even plenty of deer before the storm."

"And I thought you knew what you were talking about!" Andrew teased, still whispering.

The hunters pressed on, all the way to the lake, without seeing a sign of a deer.

"Well, guide, now what do we do?" Andrew asked.

"Boy, you've got me," Tom admitted.

About that time, they could see a figure come around a point of land, nearly a mile away. It appeared to be a man walking on the lake, following the shoreline and coming towards them.

"Let's wait and see who it is," Tom suggested.

Andrew nodded.

As they watched, whoever it was stopped several times and seemed to get down on his hands and knees wherever an ice ridge was pushed up against the shore.

"I think I know what he's doing," Tom broke the silence. "I think he's a trapper."

Tom's guess was confirmed as the man drew close enough for the boys to see him take a small animal out of a trap and put it in a pack he carried on his back. As he came closer, the boys could see it was an Indian, a young Indian, perhaps older than Andrew but younger than Tom. He was about Tom's height, had braided black hair, and wore a coat made from a red Hudson's Bay blanket. He carried an old muzzle-loader.

"Hello," Tom called when the stranger was in talking range. "How are you doing?"

"Good," came the reply. "Five mink."

"I'm Tom. He's Andrew."

"Me Noka," the young Indian replied.

"Where's your village?" Tom asked.

The Indian pointed across Sucker Bay to Ottertail Point, "There."

"Oh, Flat Mouth's village," Tom observed.

"Flat Mouth what you call, my uncle."

At this point, the Indian, who was also on snowshoes, got down on his hands and knees again and looked under a small ice heave near where the boys had been standing. "One more!" the trapper exclaimed as he pulled out a trap with a beautiful, big, black buck mink in it.

"What are the mink doing under the ice heaves?" Andrew asked. "Do they live there?"

"No," Noka replied as he re-set the trap, using a piece of fish for bait, "mink eat there. Get under ice, into water, catch fish."

"Have you seen any deer?" Tom asked.

"You hunt deer?"

"Yes," both boys answered.

"No, no see deer. Deer back in swamp, together, many in one place, warm in swamp, pack down snow for food. Want me show you?"

"Now, we'll have a real guide!" Andrew said, smiling at Tom, and then turning to Noka, added, "You betcha!"

And so with the young Indian leading the way, the trio headed back into the swamp. Although the snowshoes still helped, they were also a problem as the hunters worked their way between close-growing cedars and spruce and tried to move around or over snow-covered hummocks. In places the timber was so thick it was noticeably darker and gray moss hung from the lower, scraggly branches. Less than 500 yards into the swamp, there was a sudden commotion ahead. A couple of dozen deer exploded out of a small clearing in all directions, except towards the boys. Because of the thick growth it was impossible to get any of the running and jumping animals in their sites. When the hunters reached the clearing they found the snow beaten down. "Like a barnyard," Andrew observed.

Not only had the deer been feeding on the lower cedar boughs, but they had pawed up the snow until blades of green grass and budded little bushes were easily visible.

"Well, that didn't work out. Now what do we do?" Andrew wanted to know.

"Try again," Tom suggested.

"No," Noka explained, "We wait. Some come back to food."

"How long will that take?" Andrew pressed.

"Take time. They come. Climb trees. No smell."

The Indian then selected three of the bigger trees where the wind would blow the human odors away from the clearing. "Climb high. Don't shoot 'til three, four deer in open. Watch me for sign."

The climbing proved easy, but as time went by, sitting on a fairly small branch without moving became pure torture. Not only was the sitting painful for the hunters, but their clothes, wet with sweat, became clammy cold, a cold that seemed to penetrate through to the bone. After more than an hour, Andrew was beginning to wonder if the deer would ever come back, and he looked at Tom with raised eyebrows. Tom replied with a shrug of his shoulders, indicating that he wondered too. Both agreed, later, that if Noka hadn't been along they would have given up early on.

Noka suddenly put a finger to his lips, signalling silence, and nodded across the clearing. Cautiously, ever so carefully, a doe and her last year's fawn entered the opening. After what seemed several minutes, they began to feed. In a little while another doe showed, then two bucks, one a dandy. Andrew counted ten points.

Noka started to raise his muzzle-loader, then stopped as another doe and two fawns appeared.

It occurred to Andrew that Noka had not given directions as to who should shoot where. "What if all of us shoot at the big buck?" he wondered to himself. But he reasoned that Noka should shoot left, since he was in the far tree, that Tom should shoot at deer in the middle of the clearing, since he was in the middle tree, and that he should shoot to the right, and that was where the big buck was now feeding. Andrew's heart was pounding with excitement, pounding so loud he was afraid the deer might hear it. He worried about holding the rifle still. The big buck was actually coming closer! But where should he aim? The deer was now so close he decided to try for a neck shot. It was hard to watch both Noka and the buck without moving his head. Finally, out of the corner of his eye, he saw Noka start to raise his rifle. That was the signal. The buck was now facing him. It raised its head. It seemed to be looking up at him! Andrew's rifle was at his shoulder. He took a fine bead on the white patch below the buck's muzzle, and squeezed the trigger.

Noka and Tom fired almost at the same moment. The buck leaped backwards, falling on its side, dead! Tom dropped the second buck, an eight pointer, and Noka shot a doe. The remaining deer bolted for cover. Quickly, Andrew flipped the lever action of his 30-30 and threw a shot at a fast-escaping doe. Tom did the same, but at a different deer. So far as either boy could tell, neither animal was hit; they kept running. Because Noka had a muzzle-loader, he couldn't re-load in time for a second shot.

In seconds all hunters were on the ground checking their trophies: three beautiful deer.

"Think you hit other doe", Noka said to Andrew. "Deer go like this when you shoot." Noka humped his shoulders and drew in his stomach.

Andrew ran to where the deer had left the clearing. "Blood! Blood!" he yelled.

"Not go now," Noka warned. "Let lay down, get stiff, gut other deer first."

Meanwhile, Tom ran to where he had last seen his deer, but returned to say, "No such luck."

Andrew had a tough time not following his deer immediately. He had never hunted deer before and the possibility of getting two, and one a trophy buck, boggled his mind.

But obediently he returned to his buck, and by watching Noka and Tom he succeeded in cleaning out his own deer, and without cutting the stomach lining or urine bag.

The messy work done, Andrew asked while washing his hands with snow, "Now can we go after my other deer?"

"Not yet, wait longer," Noka advised.

"Hey, let's celebrate!" Tom suggested. "Let's have lunch."

The boys shared their sandwiches and doughnuts with Noka and talked excitedly about their success. Their lunch finished, the Indian smiled at Andrew and said, "Now."

"Do you think I got him?" Andrew asked as they started into the swamp.

"Think so," Noka replied. "Blood good color."

What Noka meant by "good color" was that the blood was a bright red and spattered across the snow; it meant the bullet had hit an artery.

Tom and Andrew were content to let Noka lead the way, but tracking would have been no problem for either of them. Blood was sprayed every few yards. No more than a hundred paces into the swamp, Noka stopped and turned around. He had a big grin from ear to ear as he pointed up ahead where the doe lay, very dead. Andrew couldn't suppress a shout of joy.

With Noka's help, they pulled the four deer out to the edge of the swamp where they could easily bring a couple of horses to take them back to camp.

Noka was invited to go with them, but he replied, "Me finish trap line. You come my village sometime."

"Tell you what," Tom suggested, "next Sunday morning, one week from today, you come to see us, then we'll walk back to your village with you, just so we can get back in time to help with supper."

"Yes, I do that, come in seven days," Noka promised, and turned towards the lake.

Andrew and Tom hurried back to camp as fast as their snowshoes would carry them. They proudly announced their success. Bruce was both pleased and impressed. He patted both on the back and said, "Good work, good work!"

Although Andrew had tried to visit the family horses, Dan and Dolly each day, this was really his first opportunity to work with them. He was pleased for the excuse. As the boys made their way back to the deer, they rode the horses where the going was good, and led them where the snow was too deep or the trees too thick.

By mid-afternoon they were on their way back, deer in tow. Everything was going just fine, when suddenly Dan snorted and reared up on his hind legs. Andrew who was riding him at the time, found himself lying on his back in a snow drift. Dolly also became excited and ran quite aways before Tom could get her under control. Obviously, something had frightened the horses. As Andrew struggled to his feet, he could see that Dan had stopped running. The two deer he was dragging had helped slow him down. Once he saw that Tom had reached Dan and had hold of his reins, he looked around to see what might have frightened them, "Was it a wolf? A lynx?" he asked himself. He saw nothing, but he was glad he still had his rifle. Then, just a few feet off the trail he saw a wisp of smoke rising out of the snow. Cautiously he approached. It seemed to be coming from a mound pulled up by the roots of a fallen tree. There was a little opening where the steam came out and Andrew carefully brushed the snow aside, making the opening larger. Then he saw it! Black fur, lots of black fur, lots of long, black fur. It was a bear, in hibernation for the winter. The steam Andrew had seen rising from the ground was the animal's breath. Although he knew the bear was in a deep sleep, he brushed more snow away ever so carefully, as though the slightest touch might awaken the beast. Finally, he could make out the head. He took careful aim behind the ear and pulled the trigger. The bear lurched and then was still.

Tom, of course, heard the shot. He tied the horses to trees and came on the run. "What is it? What did you shoot?" He asked excitedly.

"Would you believe I shot a bear?" Andrew asked.

"Nah," Tom replied in disbelief.

"Well, I did, I did! Look for yourself." Andrew pointed towards the den.

Just as Tom reached the hole and was bending over to examine the animal, the bear gave a final death lurch. Tom tried to jump back but caught his heel on a log submerged under the snow and fell flat on his back.

Andrew doubled over in laughter, finally stammering, "What a hunter! What a brave hunter!"

Tom had had it. It was bad enough that Andrew shot *two* deer and that one of them was a big buck, but a bear besides, and now he was laughing at his fall in the snow! He jumped to his feet with a roar and threw Andrew to the ground. Andrew was still laughing too hard and feeling too good to resist, even when Tom sat on his chest and began washing his face with snow. He finally had to beg for mercy.

Satisfied with his revenge, Tom let the younger boy up, asking as he pulled him to his feet, "Now what are you going to do with your bear? Those horses of yours aren't going to get near enough to pull it home."

"Maybe we can get an ox."

"Good idea," Tom acknowledged. "Let's get back to camp."

At the cookshack, Bruce expressed his delight with the added good news. "We'll have some roasts and chops, and the bear fat will make the best lard there is for frying doughnuts."

He then went with the boys to ask one of the oxen drivers, "Chicago Pete", to retrieve the bear.

The big oxen seemed to have no fear of bears and made no objection to pulling the huge beast (about 400 pounds, foreman Schmitt estimated) out of the den and back to camp where it was dressed out.

After supper, Andrew entertained the loggers who would listen with the hunting stories, but was careful not to tease Tom about his lesser success in front of the men. He knew he didn't have to. Andrew reasoned to himself, "I'll just let the facts speak for themselves." With a glad heart he basked in the attention he received from the 'jacks as he answered question after question. He especially appreciated the handshake from Nils and the pat on the back from Black Richard.

Before going to bed that night, Andrew borrowed an axe, went to the shed where the animals were hanging, and cut off his buck's ten point rack. Proudly, he nailed it to the wall above his bunk.

Just before the lights went out for the night, Andrew peered down at Tom from his upper berth and asked innocently, "Aren't you going to save your rack?"

"I'll be darned if I'll put up a rack smaller than yours!" he snorted,

"Shut up and go to sleep or I'll haul you back out in the snow and finish washing your smart-alec face!"

Andrew responded softly, "Gotcha", and rolled over and went to sleep, one happy boy.

CHAPTER IX
CHRISTMAS IN THE LOGGING CAMP

Caught up in the busy routine of logging camp, Andrew came to feel that he had lost all sense of time. He was up before dawn every morning, pumped water for coffee before he was fully awake, made the coffee, put food on the table, clanged the bell for breakfast, served more food, ate breakfast with the cooking crew, washed the dishes, helped cut out and bake cookies and doughnuts, helped bake bread, kept the fire going in the cook stove and the barrel stove, made sandwiches to go with the soup or stew for lunch, ate an early lunch himself, joined Tom on the swing dingle delivering the noon meal to the loggers in the woods, brought the dirty tin bowls and utensils back to camp and helped wash them, swept out the cookshack, chopped firewood, carried in firewood, helped prepare supper, set the table, put food on the table, clanged the dinner bell (unless Bruce chose to blow "Gabriel") waited table, had his own supper with the kitchen crew, helped wash dishes, set the tables for breakfast, and, if he wasn't too tired, visited with friends before "hitting the hay". And every day was just that busy. Only Sundays were a little more relaxed.

Having lost a sense of time, it was not surprising that Andrew was shocked to hear Bruce say the Sunday evening of the deer hunt, "Well, boys, this coming Friday is Christmas. We'll have to bake extra cookies and breads this week and prepare special meals for Christmas Eve and Christmas Day."

Andrew looked at Bruce in disbelief and asked, "Are you sure Christmas is this week?"

"Of course, I'm sure," Bruce replied with a chuckle. "Come here, look at the calendar yourself if you don't believe me!"

Andrew did look at Bruce's calendar where the chief cook had crossed out each date since logging camp had opened. "Sure enough," he had to admit, "Christmas is this week."

Tom, who had been standing nearby during the conversation, grumbled, "The supply sleds probably won't get here before Christmas because of the snow storm. That means gifts from home will be late this year. It just won't seem right, Christmas without gifts from home."

Andrew then remembered how his family had always sent his father a Christmas package those years when he was in logging camp. He recalled, too, how his father brought his mother, Louise, and himself belated Christmas gifts when he returned to the farm at the end of each winter. After all, there wasn't much that could be sent from logging camp except letters; these the supply sleds would bring back to Walker and they would be mailed from there. Speaking of letters, Andrew had been good about writing, once a week just as he had promised. And with every arrival of the supply train Andrew received letters from home. His mother always wrote the most and his father wrote the least, mostly because English was such a struggle for him and Andrew had never learned more than a few words and phrases of Swedish. Louise usually had something to say about friends their age and always made a big deal out of news about somebody getting engaged or married or who was now going with whom. The letters also gave assurance that his father's leg had healed and that all was well on the farm. But Andrew resolved that he'd better start thinking about what kinds of Christmas gifts he would buy each member of the family when the logging camp closed down for the season.

The days before Christmas became even busier than usual. But even with the extra cooking and baking, Tom and Andrew found time to haul in a huge spruce tree that nearly reached the peak of the ceiling of the dining room. They decorated it with strings of popcorn and cranberries. George Wilson, the oldest cookie, whittled a wooden star for the very tip of the tree. Everyone seemed to agree that the Christmas tree was "real pretty".

"What are we going to have for meals Christmas Eve and Christmas Day?" Andrew asked the head cook.

"We're going to try to have something special for each of the larger nationality groups in camp. For the Germans there'll be bratwursts and potato sausage, probably for Christmas Day supper. For the "Cousin Jacks" (English) we'll fix roast goose and plum pudding dessert for Christmas dinner. And for the Scandinavians there will be lutefisk on Christmas Eve and blood sausage with side pork for Christmas breakfast."

"My mouth is watering already, except for the lutefisk and blood sausage," Tom confessed.

"Before you get carried away with thoughts of how much you guys are going to eat," Bruce warned, "just remember you'll have to help fix and serve it!"

With all the activities going on, time passed even more quickly than usual. On Christmas Eve, the loggers were allowed to quit work an hour early to clean up for the special evening. That night when the meal was served, the non-Scandinavians complained good naturedly about the lutefisk (creamed codfish on boiled potatoes) and how awful it smelled, but Andrew noticed that everyone ate his share. And no one complained about the hot minced pie and cheese served for dessert.

Once the tables had been cleared, Bruce brought out dozens of candles, and when they had been lit, the kerosene lanterns were extinguished. The men who had musical instruments went to get them, about a dozen in all: violins, guitars, mouth organs, ukuleles, and a Jaw's harp. They made quite an orchestra.

Andrew's friend from home, Nils Jensen, was chosen to lead the singing; he had a beautiful bass voice. And so the men formed an informal choir, right where they sat at the tables, more than a hundred strong; and they sang Christmas carols way into the night. Andrew felt a lump in his throat as he remembered home and family and more than once wiped a tear from his eye when he thought no one was looking. But as he watched others, he saw many a tough lumberjack doing the same. The songfest came to a close with the singing of the first verse of "Silent Night", four times. It was the only verse most of the men knew, but it was their favorite carol.

Bruce and his helpers then hauled out the Christmas cookies and made pots of fresh coffee and tea. With everyone wishing each other a "Merry Christmas", they finally headed off to bed much later than usual, with an announcement from Bruce that breakfast would be one hour later. Andrew looked at Tom and said, almost in disbelief, "We can sleep in a whole hour!"

Christmas day broke crisp and clear and the kitchen crew went about their work filled with the Christmas spirit. There was no grumbling, no complaining, just lots of whistling and singing and wishing each other and all who dropped in for an early cup of coffee, "Merry Christmas!"

It was a day of fun and games, good food, and relaxation, except for the cooks, who worked harder than usual, just to make the meals even more special.

The dishes from the goose dinner and plum pudding dessert were just being cleared when Andrew thought his ears were playing tricks on him, he thought he heard sleigh bells! Then he knew he heard sleigh bells. He pushed open the kitchen door just as the supply train of four horse-drawn sleds came into view, with the drivers shouting “Merry Christmas” at the tops of their voices.

The ‘jacks piled out of their bunkhouses and returned the greeting. Every man suspected, and they were not disappointed, that the sleds contained more than just logging camp supplies. They also contained gifts from home. It seemed like hours to Andrew before his name was finally called and he received his package along with a lengthy letter from his family. Each gift was wrapped separately. There was a rawhide belt along with a knife and sheath his father had made, wool mittens and a scarf knitted by his sister, and two wool shirts tailored by his mother. There was also a hand-knit sweater from his Aunt Emma in St. Cloud, which proved to be a little small; she hadn’t realized how much he had grown. And there were six books, each from a different friend or relative. Andrew’s mother explained in the letter that she suggested sending a book whenever anyone asked what they could give him for Christmas. “Read them all before you come home, Andrew,” she admonished. “It will help you continue your education.”

Tom had faired equally well and the boys enjoyed viewing each others gifts.

“I wish I had something to give you, Tom,” Andrew said sincerely.

“Ya, same here,” Tom replied, a little sadly, then added, brightening, “I’ve got an idea! Your sweater is a little small, the one I got is a little big, so let’s give each other a sweater!”

“You betcha!” Andrew responded, and the exchange was made. The original giver of each sweater would never know how she made each boy doubly happy.

CHAPTER X
FLAT MOUTH'S VILLAGE

It was the Sunday morning after Christmas, and Andrew and Tom were still working on the breakfast dishes when Noka, the Indian lad from the village on Ottertail Point, came to the cook shanty, just as he said he would. In the excitement of the holiday season the boys had forgotten their invitation to Noka and their promise to return with him to his village, so they hadn't asked Bruce for time off. After making their Indian friend welcome, Tom approached the head cook, but while he was still thinking of what to say, Bruce observed, "See you got company."

"Ya," Tom replied, "we wanted to talk to you about that."

"Well, talk to me."

"We met him when we were deer hunting last Sunday and invited him to camp."

"So?"

"Well," Tom added apologetically, "we promised to return with him to his village, Flat Mouth's village, for dinner today."

"And now you want the rest of the day off," Bruce guessed.

"Ya, and in all the extra work getting ready for Christmas we forgot to ask, but we could be back in time to help with supper."

"But you missed the noon meal last Sunday when you went hunting."

"That's true, but we did bring back four deer and a bear. Just think what a mess we'd have been in if the supply train hadn't gotten here. And going hunting was your idea."

"I don't know," Bruce hesitated. "Finding volunteers to help cook and clean up on the men's day off is never easy."

"It's really kind of important," Andrew butted in. "You see, it was Noka who showed us where the deer were and how to hunt them, and we promised—."

Tom elbowed Andrew in the ribs.

"Oh?" Bruce expressed surprise. "Now we hear the *rest* of the story. And you had the whole camp thinking you were such great hunters," he teased.

Andrew and Tom looked down in embarrassment.

"All right. I'll check around for helpers. Show your friend around. Stop back in a half hour, but don't leave camp without seeing how I come out."

Once outside the door, Noka asked, "What you do in logging camp? You cook?"

Tom and Andrew could tell by the way he asked the question that he knew what they did but couldn't believe it. Finally Andrew responded, "Ya, so what?"

Noka smiled, said nothing for a minute, and then answered, "In Indian village, only *woman* cook."

"That's true in our homes, too," Tom said defensively. He then tried to justify their doing "women's work" by adding, "In a logging camp with over 100 hungry men to feed, women could never handle all the work."

Eager to drop the subject, the "cookies" quickly led Noka around camp, from building to building, explaining the logging operation with great authority, trying to impress him with their knowledge. Near the end of the tour, they met Bruce coming out of one of the bunkhouses. "Well?" Tom asked, wondering if they could go.

"I guess it'll be all right. I got Hank Benson to help and he thinks he can find someone else. Just be sure you're back before dark to help with supper."

"Thanks, Bruce," both boys said at once. Then, as an afterthought Tom asked, "Can we borrow the rifles?"

"Why?" Bruce wanted to know.

"We might see some game."

"My guess is you're *afraid* of game!" Bruce teased, but agreed. "Ya, go ahead."

It was nearly a two-hour hike to the village. They used snowshoes as far as Sucker Bay, but didn't need them on the lake; strong winds since the storm had blown the ice fairly clear of snow. What was left had hardened, making walking relatively easy.

The white boys were full of questions and gave Noka a real workout with his limited English vocabulary. Of course, one of the first questions they asked was, "Is Chief Bug back on the lake?"

"No," Noka assured them, "Army gone, Chief Bug not back." He then added with a smile, "You plenty safe with me."

As the travelers approached Ottertail Point they could hear sounds from the busy village and smell food being prepared. They were first met by the village dogs who barked quite ferociously at the strange smelling whites. The dogs were scary; they looked like wolves. The fact was that some of them weren't many generations removed from their wolf ancestors. Andrew and Tom were glad they arrived under Noka's protection.

The village was larger than the boys had expected, more than 200 people. Several log cabins were scattered throughout the clearing, but most folks lived in lodges covered with bark and animal skins, much as their ancestors had for generations. It was a relatively warm day for midwinter and many of the residents of the village were outside. A number of children approached them as they left the lake, anxious to get a closer look at the strangers with white skin and round eyes.

"I've never been stared at so much in my life," Andrew whispered to Tom.

Two of the older children, a girl of about twelve and a boy maybe a couple of years older, fell in step alongside the travelers. "My brother and sister," Noka offered. "Him Wabose, her Winnetka."

Both seemed quite shy and just nodded and smiled when the boys spoke to them. "Not know much English," Noka explained.

The village was a busy place. Considerable cooking was going on out-of-doors; trappers were skinning out animals such as beaver, mink, and marten; firewood was being hauled and cut; children were laughing and shouting as they played their games; older people were visiting; everyone was doing something.

A number of Indians, both men and women, entered the village from the opposite side of the point, pulling sleds full of fish, mostly whitefish, walleyes, and northerns, which they had caught in their under-the-ice gill nets. Noka identified one of the fishermen as his father, Majigobo, and he introduced his guests to him. Majigobo smiled and said, "Happy you come," and led the way to the cabin.

Noka's mother, Ozha, met them at the door with a smile and a barrage of Ojibway words.

"Mother speak no English," Noka explained, "but she say you welcome."

The cabin was small compared to the houses Andrew and Tom were used to, just one big room, but it was warm and comfortable. There was a cast-iron stove that served both as a heater and as a place for indoor cooking. There were no cupboards or closets, just some open shelves, but everything seemed to have its place. Much was hung from the walls and ceiling, everything from tools to clothes to dried foods and herbs to guns. Blankets on fresh spruce boughs served as beds as well as places to sit. It was truly a home.

Noka spoke, "We have special feast today. It is called Ostenetahgawin, feast of first fruits. It is in honor of my brother, Wabose; him trap his first beaver. It is being roasted out in back of cabin. Very good to eat. Come see."

Noka led the way out back. There was a very large beaver carcass suspended on a spit over hot coals. Other foods were cooking in kettles.

Andrew and Tom weren't so sure just how good the beaver would taste, but when meal time came, they were pleasantly surprised. Tom described the beaver as tasting like beef. They also enjoyed the fish broth and wild rice served as a vegetable. The bread was flat and tasted very different. When asked what it was made of, Noka pointed at a nearby oak tree and said, "Acorns."

After dinner, Andrew asked Noka about Chief Flat Mouth. "Did you tell us when we first met that Chief Flat Mouth is your uncle?"

"Thats right," Noka replied.

"I understand he is a great storyteller as well as a great chief. Do you suppose we could meet him?"

"Can do," Noka answered. "Come."

Andrew and Tom followed their friend to the largest cabin in the clearing, a cabin with real glass windows. As they approached, the chief himself stepped out of the door. In deference to the white boys, he greeted his nephew in English, "Noka, you bring friends to my house. They welcome."

Noka responded by introducing Tom and Andrew.

"Before you come in cabin, I dress to receive my guests. Wait here." With that, the chief smiled at the boys, turned, re-entered the cabin, and closed the door behind him.

"What is that all about?" a puzzled Tom asked of Noka.

"By putting on special clothes, my uncle pay you same honor as other important white men come see him."

Minutes seemed like hours as the boys waited. Each expected the chief to reappear wearing a feathered headdress and elaborately beaded buckskins. Imagine then their surprise when the door finally opened and

Chief Flat Mouth emerged in a blue serge suit with vest to match and a high silk hat. A huge silver medallion hung from his neck, suspended by a wide blue ribbon.

"Now, come in please."

In a state of shock, Tom and Andrew followed Noka and his uncle into the cabin.

"You pay us a high honor to dress up for us," Tom stammered.

"These same clothes my father, also called Chief Flat Mouth, wear to Washington when he signed treaties and met great White Father."

"He met the President of the United States?" Andrew asked in surprise.

"Yes, President Pierce. He appear before Congress with other Indian chiefs. Statue of my father in Washington."*

Andrew and Tom were properly impressed.

"Tell about medal," Noka prompted.

The chief fondled the medal, looking down at it as he spoke, "Lieutenant Pike give this to Father when he come this lake."

"That must be Lieutenant Zebulon Pike," Tom whispered to Andrew. "Pike's Bay of Cass Lake and Pike's Peak in Colorado are named for him."

The chief then seated himself on a huge wooden trunk. "Like a king on a throne," Andrew would say to Tom on their way home.

The chief motioned the boys toward some chairs by a table. The white boys were surprised to see not only a table and chairs, but other "whiteman's furniture", even a bed.

"Why you boys come my village?" the Chief asked.

Tom explained, "We work in the logging camp across the bay. We met Noka when he was trapping over there and he invited us to dinner."

Noka added, "Wabose have Ostenetahgawin feast; he get his first beaver."

"Good," Flat Mouth replied, "After we talk, I go honor nephew and have some beaver to eat if some left."

"It was very good," Andrew said, making small talk.

"We especially wanted to meet you, sir," Tom added, "The men at the logging camp have told us that when you have visited there, you have been a great storyteller and know much about the history of your people."

**The statue of Chief Flat Mouth is still on display in the U.S. Capitol Building. Chief Flat Mouth II visited Washington D.C., himself, in 1907.*

Without waiting for the Chief to reply, Andrew added, "Yes, have the Pillager Indians always lived on this lake?"

"Pillager Indians part of Ojibway Nation," Flat Mouth explained, "Me fourth generation on lake. My people once live on ocean you call Atlantic. Iroquois with guns kill many my people. My people had no guns. My people come here, but stop long time on Kitchi Lakes."

"Kitchi means great," Noka explained.

"Like Lake Superior?" Andrew asked.

"Yes," both Indians replied.

"Were the Ojibway the first Indians to live on this lake?" Andrew asked.

"No," replied the chief. "Sioux here before Ojibway. Others before Sioux. We capture lake with much fighting. Many die."

"For many years," Noka added, "Sioux try to take back lake. We always win."

"Who tell story?" Flat Mouth asked his nephew a bit sternly, "Me or you?"

Noka apologized in Ojibway.

"Have the Ojibway and the Sioux always been enemies?" It was Tom's turn to ask a question.

"No," the chief replied. "My grandfather, Chief Yellow Hair, him tell of time when Ojibway and Sioux people go to Crow Wing River every winter and hunt and trap side by side, but that not last long."

"On my way up to logging camp, I heard a story of how a band of Pillagers defeated a small army of Sioux where the Crow Wing River goes into the Mississippi. Would you tell us another story of Indian fighting?"

"You speak of Crow Wing," the chief responded. "My father tell of time when he small boy. Many from this village spend winter hunting and trapping for French traders where Partridge River meet Crow Wing. Frenchman called Blacksmith had trading post there."

"Why did your people leave Leech Lake and go to the Crow Wing in the winter?" Tom interrupted.

"In those days, very hard hunt and trap here in winter. All forest. Not much game. Not always trader here to trade with for what my people need. Many beaver by Crow Wing."

"So what happened at the mouth of the Partridge?" Andrew brought the chief back to his story.

"One day, small band Pillagers find trail of many Sioux in snow. Know them Sioux by smell of different tobacco in air. Sioux heading for small Pillager camp. Afraid all be killed. Pillager braves run past Sioux and make trail Sioux sure see and follow away from camp. Pillagers lead Sioux to trading post at Partridge, built like fort. Warn Frenchmen and other Pillagers. Everybody go inside fort. Sioux come. Maybe 200. Find Fort. Cross river and prepare attack. Put on war paint. Have war dance. But Sioux only have few rifles, mostly bows and arrows. Frenchmen and Pillagers all have rifles."

In their minds, the boys could easily see and hear the preparations for battle.

"When ready, Sioux give war whoops and charge cross river. Pillagers and Frenchmen shoot. Kill many. Sioux go back. Sioux charge again. More killed. Sioux go back. Sioux stay other side. Shoot arrows high in air so fall in fort."

The chief made a motion with his hand, tracing the arrows' flight in a high arc, mortar style.

"Some wounded. Nobody killed. Pillagers and Frenchmen hide better. Pretty soon Sioux all out arrows, no dare charge. Before go away, Sioux cut holes in ice; put dead in river. They go away. Come back no more."

"Wow!" Tom said out loud, then added, "Why would they put the dead warriors in the river, under the ice?"

"Sioux 'fraid Pillagers scalp dead warriors. Believe that how they look in Happy Hunting Ground, no hair!"

With Tom and Andrew's encouragement, the chief told stories of battles west of the Red Lakes, on the Mississippi, on Battle Lake, and more. The chief indeed proved to be a great storyteller, and the boys were equally good listeners. It was Tom who first noticed that the cabin was growing darker and that the afternoon was far spent. He reluctantly said to Andrew, "We'd better get started for camp or Bruce will scalp us!"

Andrew and Tom thanked Chief Flat Mouth and Noka walked his guests as far as the lake.

"Come see us and stay longer," Tom urged Noka.

"For sure," Andrew added.

With a final wave, the boys headed across Sucker Bay. By now the sun had slipped below the horizon and twilight was setting in. As the boys neared the far shore, a wolf howled to the west on Ricebed Point. Another wolf answered up the shore to the right. Then another and another let go on the point to the left. The boys stopped. "I think the hair on the back of my neck is standing up!" Tom said with a touch of fear in his voice.

"I've got goose bumps all over," Andrew admitted. "I think those wolves are talking about us!"

It was then that a small pack of wolves, seven in all, became visible on the ice as they left the wooded shoreline of the point. And they headed right for Tom and Andrew.

"What should we do? Should we run for it?" Andrew asked urgently.

"No, we've got our rifles if they do attack. Besides, we could never outrun a pack of wolves. Just keep walking."

"Do you think they will attack?"

"I really don't think so. My guess is that with the wind the way it is they can't smell us and they think we're a couple of deer."

Tom was exactly right, but the wolves kept coming at a full run and it was very unnerving. Still a half block from shore, Tom suggested, "They're getting too close for comfort. Let's sit down on the ice and use our knees to steady our rifles. But don't shoot until we have a good chance of hitting one."

As the boys sat down, the wolves realized their intended prey weren't acting like deer. They slowed their approach but kept coming. At about 200 yards, Tom instructed, "All right, aim high, let 'em have it!"

The rifles cracked; the silence of the wilderness amplifying the sound. The wolf pack turned and ran back towards the point with each boy firing a parting shot for good measure.

"It worked!" Andrew yelped in relief. "But I wish we'd got a couple."

"I'm just glad we scared them off. Let's get out of here."

The boys kept looking over their shoulders as they snowshoed the rest of the way through the darkening forest.

Upon reaching the cookshack they flung open the door and both began talking excitedly at the same time.

"You won't believe it, Bruce!"

"We were attacked by wolves!"

"But we fought 'em off."

"Sure, sure," Bruce replied, "You are just grasping for excuses for being late."

"No, honest, Bruce," Andrew said defensively, "If it hadn't been for your rifles we wouldn't be here."

It wasn't until the crew was cleaning up the supper dishes that Andrew and Tom were finally able to convince the others of the truth of their adventure.

CHAPTER XI
A REAL LUMBERJACK AT LAST

Hard work continued to help make the time pass quickly. As January blended into February, Andrew noticed that there were more hours of daylight. The added exposure of the earth to the sun, however, didn't seem to warm things up. In fact, just the opposite appeared to be the case. As Ole Olson observed in his thick Norwegian accent, "As the days grow longer the cold grows stronger." There were many days it was so cold Andrew was grateful for his inside job. Driving the swing dingle on the lunch rounds each noon gave him all the fresh air he wanted. On those nights when the temperature plummeted to 30^0 to 40^0 below zero, it wasn't unusual to be awakened by the boom of sap or moisture exploding in nearby trees or in the logs of the building itself.

It was nearly March before the first long-awaited thaw came to the north woods. Noon temperatures were actually above freezing for a change and the waist-deep snow began to settle. Unfortunately, the warmer weather brought with it a really serious problem—a flu epidemic. No one knew whether the germs arrived by supply train or whether the warmer weather thawed out some flu bugs lurking in camp. Either way, the disease hit with a vengeance.

Andrew and Tom were caught in the first cycle. Both boys woke up one morning with upset stomachs, diarrhea, and every bone and muscle in their bodies aching. As they compared ailments, Andrew asked Tom, "Are you sure somebody didn't beat me up last night? I hurt all over."

Bruce mercifully encouraged the boys to stay in bed and explained, "I don't want you food handlers to spread the disease all over camp."

As Foreman Schmitt arrived for breakfast, he reported that several of the men had checked in sick and mused to the cook, "If this bug spreads through camp, we're going to fall behind schedule."

The Bull's fears were to be realized. During the next 30 days, scarcely a man would escape the ailment. Worst of all, it was a very severe variety of the flu. Even after a person stopped vomiting and running to the outhouse, he would be too weak and sore to work for many days; but from Andrew's perspective, once again an ill wind was to blow some good his way. Being young and normally in robust health, he and Tom were on their feet and back to work in less than a week. Meanwhile, a dozen or more new lumberjacks would report in sick each morning. Foreman Schmitt grew more upset daily and grumbled non-stop about how far behind the camp was getting in its work. "We'll never make our quota by break-up," he worried again and again.

It was after one of those pessimistic outbursts that Andrew sat down by the boss as he was dunking doughnuts late one night and said with a smile, "Hey, Mr. Schmitt, be careful there how you dunk your doughnut. Bruce says you aren't supposed to get your thumb or fingers wet beyond the first knuckle!"

"Don't try to cheer me up, Andrew. The situation is too serious."

"Well, I do have an idea that could help a little."

"What's that?"

"I don't know if you realize it, but Tom and I have had a lot of experience helping our fathers cut trees at home. I know I've helped cut a couple of hundred myself. With so many men sick and not eating, Bruce probably could get by without us for awhile."

"I don't know, Andrew, you're pretty young. It's hard work and it's dangerous work."

"Golly, Mr. Schmitt, would there be any harm in letting us try? We'd be careful, and if we couldn't handle the work we'd go back to the pots and pans."

At this point, the foreman turned towards the kitchen where the head cook was setting out some things for the next morning's breakfast and shouted, "Hey, Bruce, got a minute?" The big man poured himself a cup of coffee, then walked over and sat down across from Andrew and the "big boss", and asked, "What's on your mind, Mr. Schmitt?"

"Well, Andrew here has this crazy idea that he and Tom could help out in the woods while so many men are sick. I guess every little bit helps. Do you think you and your other cookies could get along without them while you have fewer mouths to feed?"

"It would be a pleasure just to get them out from under my feet," the cook replied with a wink at Andrew.

"All right!" Andrew shouted.

"Understand, now," Schmitt warned, "You and Tom will each be assigned to an experienced 'jack and I expect you to do exactly what they tell you."

"Of course," Andrew assured him.

"And if you boys do a good job, I'll raise your pay to a full dollar a day while you're in the woods."

"Sounds good," Andrew replied and excused himself to tell his buddy the news. Tom was already asleep on his bunk, so Andrew enjoyed shaking him awake and then when he was conscious asked, "How would you like a raise of twenty-five cents a day?"

"Huh?" was all Tom could say.

"Mr. Schmitt has agreed to let you and me work in the woods while so many men are sick. We'll be real honest-to-goodness lumberjacks!"

"You're kidding."

"No, it's for real."

"Whose idea was it?"

"Mine," Andrew answered, proudly.

"Way to go! I guess I owe you one."

"You got that right, Tom, and I won't let you forget it."

Only the realization that they would need a full night's sleep kept the boys from talking much longer as they speculated on what their new jobs would be like.

The next morning was one of the very few times Andrew and Tom were awake before Bruce bellowed, "Daylight in the swamp!" By the time the men began coming in for breakfast, both boys were dressed for work, except for their mackinaws and caps, and were seated at the table waiting to be served. Rarely have two young men been the recipients of more teasing, especially from the other cookies, but they loved every minute of it. At the end of the meal foreman Schmitt advised the boys, "Tom, you'll be assigned to Ole Olson, and Andrew, you go with Nils Jensen."

Andrew was delighted. He liked and respected his neighbor from home and was confident he would be treated well.

As they headed into the woods, Nils observed, "If your pa could see you now, he'd be proud as punch!"

Andrew thought he knew what logging was all about. After all, he'd helped his father cut down many trees at home and he had watched the lumberjacks ply their trade numerous times. But he didn't realize how hard the work was, being on one end of a heavy cross-cut saw from daylight to dark. About the only breaks he had came while swinging a double-bitted axe to trim the branches off the fallen trees. This was a job normally left to a separate crew of less experienced 'jacks, but with so many men sick, each cutting team did their own trimming.

When he saw Tom at lunch and asked how it was going, his friend replied, "Great, just love it! How about you?"

"Oh, ya, really enjoying it," Andrew lied. Actually, he had never worked so hard in his life and wondered if he could last through the day.

George Wilson brought lunch in the swing dingle and asked, "Are you guys ready to come back to the kitchen?"

Both boys put on a good front and assured him:

"No way! This is the life."

"This is a lot more fun."

Although the work was hard, it was also interesting and satisfying. Andrew really got a kick out of yelling,"T - I - M - M - B - E - R - R - R ! each time a tree was about to fall. He tried to improve his skills by paying close attention to how and where Nils started each cut and how he allowed for the wind, so that each tree would fall exactly where it was supposed to. He made sure that the tree would not get hung up on another tree on the way down. He also pointed out how important it was that it not fall over a big limb of another tree which then might act as a pivot or fulcrum, raising the stump end of the tree up in the air where it could fall on one of the cutters. Care was also taken to notch the backside of the tree so it wouldn't kick back when it fell, again a threat to the cutters.

From time to time, Nils asked Andrew how he was doing or if he wanted a rest, but Andrew was too proud to acknowledge the pain or how tired he was. Determination and plain Scandinavian stubborness helped him through the day. Quitting time came none too soon and Andrew was grateful for a ride back to camp on an empty bobsled.

Both Tom and Andrew had difficulty keeping their eyes open during supper and as soon as the last 'jack left the table, they dove for their bunks.

"So how do you feel, Tom?" Andrew asked as he undressed.

"Never better. Great day."

"You're a liar!" Andrew responded, half laughing.

"Ya, I know, I just hurt all over. You?"

"Just awful. I think I'm going to die. Good night."

The next morning, Bruce not only had to shout twice, but he also had to give both Andrew and Tom a good shaking to arouse them. As they dressed, the boys made honest confessions to each other about their aches and pains; however, each truly looked forward to the new day in the realization that they were actually working as adults. After all, they were now real lumberjacks.

It would be more than three weeks before enough 'jacks would return to good enough health to go back to work and thereby make sufficient demands on the cookshack to make it necessary for the boys to return to their duties in the kitchen. These would be three weeks of excitement, satisfaction, and rewarding hard work. Muscles would be hardened, stamina would be developed. By the end of the first week, the boys were able to finish a day's work without feeling exhausted. Although they welcomed a day off on Sunday, they genuinely looked forward to returning to their jobs in the woods on Monday morning. Best of all were foreman Schmitt's words of praise when he told them, "I'm satisfied you have earned your raise."

Andrew had been a little on the chubby side for as long as he could remember, and being within arm's reach of so much good food all winter had added pounds in the wrong places, but the hard physical labor was making a difference. Each week it gave him great satisfaction to have to cut a new hole in his belt. One night after dinner, he bragged to Bruce, "I want you to know I've tightened my belt three notches!"

The big cook looked him straight in the eye and said without smiling, "Gee, that must have hurt!"

In many ways, the three weeks in the woods were the highlight of the winter for both Andrew and Tom, except for an unfortunate accident. It happened during the second week. A crew was loading a bobsled with especially large logs. This was done by using an ox or a team of horses to pull one log at a time up in place. The log was cradled in a chain and as the animals moved forward on the other side of the sled, the log was pulled up into place on a slide formed by two smaller logs. As the last log was being lifted onto the top of the load, one of the slide logs cracked and the chain broke. It started an avalanche and in spite of the warning from others, one lumberjack was caught in the way. He wasn't quite able to scramble to safety. Life was literally crushed out of him. Andrew was working nearby and when he heard the screams and the crash he rushed to the scene. He helped pull the logs away, but it was too late. The man was dead. It was Boomer Smith.

All the fear, all the hatred Andrew harbored for the bully melted away as he viewed the limp, battered body. Sadly, he helped escort the remains back to camp: Foreman Schmitt announced there was little choice but to bury him at the campsite. He ordered a fire to be built to thaw out the ground so a grave could be dug. It would take a couple of days. Andrew approached the foreman with a request: "I'd like to volunteer to keep the fire going the first night."

"Sure, Andrew, I guess that would be all right; but I must say, I'm a little surprised after the way he treated you."

"Ya, I know, but I kind of feel responsible for the beating he took from Black Richard, I guess I'd feel better if I could just do something for him."

And so, following brief graveside services led by the foreman, Boomer Smith was buried in the Minnesota woods more than a thousand miles from his native Maine. His loggers' boots were tied together by their laces and hung from the crude cross that marked his final resting place.

CHAPTER XII
THE BIG BOSS COMES TO CAMP

March winds brought warm weather, melting most of the snow, which made it very difficult to haul logs to the lake, where they would be floated to the mill at Walker come break-up. It gave foreman Schmitt something new to worry about. "How can we meet our quota if we can't use our bobsleds to haul logs? Skidding a few logs at a time behind a horse or ox takes forever," he complained to anyone who would listen.

The first supply train in March came early, taking advantage of the last remaining snow and ice. Once the snow and ice were gone, they would have to use wagons and that would mean added problems as the run-off and spring rains turned the trail into mud. On that March supply train came "Mr. Excitement", the big boss himself, Tomas Barlow Walker, owner of the camp and head of vast logging operations in north-central Minnesota. John Schmitt wasn't expecting Mr. Walker quite so soon, and he was so excited Andrew thought he might have a stroke! Andrew was pumping water just outside the cookshack when the foreman met the arriving owner. He had always thought the title the men had pinned on the foreman, "Bull of the Woods", was truly appropriate. He was very much in charge of every situation and his word was final. No one ever argued with John Schmitt. So it came as a surprise to Andrew when big John bowed and scraped before Mr. Walker like he was royalty, behaving more like a wimp than the top man in camp. He greeted Mr. Walker by nearly shaking his arm off and immediately began talking about the camp's problems and making excuses.

"Mr. Walker, I'm afraid you're going to be disappointed in our production. I don't see how we can make our quota. We were hit so hard by the flu, you know, and now, just when we're starting to get caught up, we have this early spring that's really going to slow us down."

"How far behind are you?" Mr. Walker asked, growing concerned because of his foreman's obvious anxiety.

"Well, we're about a thousand logs below our goal for the winter."

"John, John," Mr. Walker said calmly, shaking his head, "You worry too much. I haven't seen a March in Minnesota yet without snow, and you really aren't that far off target. If I were a betting man, and you know I'm not, I'd bet you make your quota."

"I wouldn't bet against you, Mr. Walker, even if you were a gambling man, and I know you're not, and I wouldn't want to bet against this camp or my men, but I'm afraid I'd have a pretty safe bet."

"Well, John, we'll have to wait and see. Time will tell. But I know you well enough to have every confidence you and your men have done the best you can, and that is all I ask."

"Thank you, Mr. Walker, thank you for your confidence," the foreman muttered, again taking his employer's hand and nearly shaking it off.

Andrew stood in awe, amazed that anyone could have this man of iron, whom he admired so much, so shook-up; but when he shared his surprise with Bruce, the cook explained, "Remember, Andrew, Mr. Walker is our employer and he could fire anyone of us—you, me, or even Mr. Schmitt, for any reason, even if he didn't like the way we parted our hair. He is an important person. He is head of a mighty logging operation. Besides, do you know any other living person who has a town named after him?"

"No, I guess not," Andrew admitted, "but if Mr. Schmitt is so afraid of Mr. Walker, I don't know how I should act if he says anything to me."

"Just be yourself. Mr. Walker is really a fine man."

"Aren't you afraid of him, Bruce?"

"Not really afraid, but I do have a great deal of respect for him; and just to be on the safe side, we'll fix his favorite foods while he is here."

"What are they?"

"Well, some of them are pot roast with brown onion gravy, chicken and dumplings, bean with bacon soup, wild game mulligan stew, and his favorite dessert is apple pie. I'll make the crust with the lard I rendered from the fat of that bear you shot. And I nearly forgot, he really likes my Swedish rye bread. I usually send a few loaves with him when he leaves camp."

"Do you have any advice, Bruce, on how to get along with him; that is, if he talks to me or if I get to serve him?"

"Like I said, be yourself, but he is a strict Methodist. He doesn't go for bad language or gambling, and he's death on whiskey. He's really religious."

That evening, following Bruce's pot roast supper, Mr. Walker addressed the men. Andrew thought he was a fair and reasonable man as he said, among other things, "I know you are all concerned about meeting the production goal for this camp. So am I. Demands for lumber are really growing as this state grows, particularly Minneapolis and St. Paul. Homes and stores and schools and churches are going up every day. With that kind of demand, prices are good. I know you've had some tough luck: first the flu and now the snow is going fast, but I've been telling John Schmitt here that I think you can still make your quota, especially if we get some more snow. You really aren't that far behind. I know you are going to do your best, but as a special incentive, I'll give every man three days extra pay if you reach your goal!"

Spontaneous cheers and applause filled the dining hall and brought the men to their feet.

"We'll do it for you, Mr. Walker!" foreman Schmitt promised, again about shaking his employer's arm out of its socket.

The next day was Sunday and the men were on their best behavior. The sky pilot, Reverend Frank Higgins, knew Mr. Walker would be in camp and showed up in time for the noon meal. That evening, every man attended the church service, even Black Richard. Pastor Higgins called on Mr. Walker to read the scripture and lead in prayer. It was the longest prayer Andrew could remember hearing, but it left no doubt about the sincerity and depth of the man's spiritual convictions. And the singing was never better.

On Monday, Mr. Walker carried out a full inspection of the camp and the logging operations and then reviewed the year's work and a few plans for the next year with foreman Schmitt. He even had a visit with Bruce about the cooking operations, after which he told the kitchen crew, "You fellows are really important to the success of this camp. You may have heard it said that an army moves on its stomach. That's even more true in a logging camp. I've found over the years that the best way to keep a 'jack happy and working his best is to give him plenty of good food."

Mr. Walker then added as an afterthought, "My goodness, Bruce, I nearly forgot. I need to steal one of your cookies to help out on the first Wannigan that will follow the log drive down the Crow Wing. I promised George Salisbury, the cook, you know George, I'd bring an experienced helper back with me."

Andrew's heart skipped a beat. And he noticed that Walker didn't ask Bruce if he could spare someone, he just said he was taking a cookie back with him to the Crow Wing River.

Before Andrew even had time to wonder who it might be, Bruce responded, "Take Andrew, here. He lives near Little Falls and can go home from there. He's a good worker, even worked in the woods along with his buddy, Tom, when so many men were sick." Bruce put his hand on the boy's shoulder as he talked.

"Good," Walker responded, then turning to Andrew, added, "I've had my eye on you, young man. I like the way you hustle and I like the way you sing. You've got a good voice. I was impressed with the way you knew all those hymns by heart last night, never had to use a songbook. Are you a Methodist?"

"Thank you, sir. No sir, I go to the Swedish Covenant church in Upsala."

"Well, that's almost as good!" Walker answered with a chuckle.

Andrew noticed that no one, neither Mr. Walker nor Bruce, had asked him if he wanted to leave camp early. It was decided and that was that. He rather resented having no voice in the matter; but, on the other hand, very much liked the idea of traveling with the "big boss", as Mr. Walker was sometimes called behind his back.

"When do we leave, sir?" Andrew asked.

"First thing tomorrow morning. Maybe Bruce will let you off breakfast duty."

"Sure thing, Mr. Walker," Bruce agreed.

"I'll have to see if Nils Jensen will take our horses home with him," Andrew thought out loud.

"Are they good riding horses?" Mr. Walker asked.

"The best, sir!"

"Good, we'll take them. I was going to arrange for horses for us anyway. I'll see to it that someone takes them to your farm from Akeley."

"May I ask where Akeley is?" Andrew asked.

"You mean you haven't heard about my new town? I've started a new mill and a new community at the head of the Crow Wing Lakes. I've named it "Akeley" after Henry Akeley, one of my partners. It's about ten miles west of Walker."

"Oh," Andrew answered, embarrassed that he hadn't known.

"I guess I surprised a lot of folks by choosing a new site for the expansion instead of adding to the mill in the town named for me. Frankly, some of the merchants are in a tizzy because I didn't expand there, but I'm disgusted with that town. If I could, I'd take my name away. I don't like being associated with any place that has so many taverns and brothels, with their "ladies of the evening", although in Walker they really aren't ladies of the evening, they're ladies of any old time of day!"

Walker stopped, but when no one had anything to say, he went on, "I'll tell you one thing, there's not going to be any liquor bought, sold, or consumed in my new town. It's going to say right in the deed of any property I sell that anybody who has liquor on the premises will lose that property!"*

It took Andrew only a matter of minutes to pack his belongings. The only things he brought home that he had not brought with him to camp were a few souvenirs like his buck horns. He spent the last evening saying good-bye to his friends. As he shook hands with Black Richard, the big man advised, "Stay out of trouble, lad, but if anyone bothers you, tell 'em Black Richard is looking out for you."

Next morning, saying good-bye to his buddy, Tom, was the hardest; but since the latter lived north of St. Cloud and less than 30 miles from Upsala, they agreed to get together at least a couple of times during the summer. As they parted, Tom confessed, "I guess I'm a little jealous of you traveling with Mr. Walker. It's really quite an opportunity; don't blow it!"

Andrew's last words to Tom were, "Next time I see you I'll be ready to take you on in a wrestling match. I've got a lot of getting even on my mind!"

"I know you're getting bigger," Tom responded, "but that doesn't mean you're getting better. There's no way you'll ever take me!"

Shortly after breakfast, Walker and Andrew mounted Dan and Dolly and were on their way to Akeley, by way of Walker.

The trail to Walker was a good one, and with the snow about gone, travel was easy. As the town came into view, Mr. Walker again derided his namesake: "Just look at all the bars. I think if you count them, Andrew, you'll find there are more than forty. "Sin city", that's what I call

---

**Walker made good on his threat but in later years the courts ruled the provision illegal.*

it. I hate to think how much of the money I pay the men who work here ends up in the pockets of the barkeeps or in the hands of loose women. We'll leave for Akeley tomorrow just as soon as I tend to some business at the mill."

They stayed over at the local hotel; it was Andrew's first such experience.

It was late afternoon before the "captain of industry" and his young traveling companion headed out of town. Walker confided to Andrew, "I'm tempted to stay over another night. I don't feel very well, probably caught that flu bug while I was in your camp; but if I'm going to be sick I don't want to be laid up in this hell-hole."

There were snowflakes in the air as the travelers headed west. Walker observed, "Maybe John will get the snow he needs, just like I predicted."

Northern Minnesota can usually count on at least one good snow storm in March or even April, and that's just what hit, in full fury, about the time Andrew and Walker reached the half-way point to Akeley. A heavy, wet snow was whipped by gale-like winds, causing near zero visibility, and it became very difficult to see the trail. Finally, Walker reined up his horse and admitted, "I think we'd better hole-up here, we're not going to make it. I guess I should tell you, too, son, I feel pretty sick. I've got the flu and I've got it bad. I'm shaking all over and I'm chilled to the bone."

Andrew was alarmed, but he tried to appear cool and in control as he suggested, "Wrap yourself in blankets, Mr. Walker, while I build a fire and make a shelter."

With that, he dismounted and began collecting a mound of birchbark, particularly the inner layer which would still be dry. In minutes he had a roaring fire going. Walker acknowledged, "Good work. Thank you. We'll make it."

Andrew then cut dozens of spruce boughs. He bent some sapplings over and tied them in place to make the framework of a hut. He covered the top and sides towards the wind with the boughs. He built a new fire at the mouth of the hut and put a heap of additional boughs on the ground under the shelter for them to lie on. After making Walker comfortable, he gathered a huge pile of firewood, enough to at least last through the night. Fortunately, it was not real cold, just wet and miserable.

Walker was feeling the full effects of the flu with all its unpleasant side effects, but by putting on his spare clothing and covering up with a couple of blankets he finally stopped shivering.

"I'm very thankful you are with me, Andrew, you're doing a fine job. Thank you."

"Don't worry, Mr. Walker, you're going to be all right."

It was a long night for Andrew. He slept fitfully and spent his waking hours tending the fire. The wind continued to howl and the snow never let up, but the spruce bough shelter and the fire kept the travelers dry and reasonably comfortable. Morning brought no change in the weather. When Mr. Walker opened his eyes, Andrew asked, "How are you doing?"

"I'd like to say I feel better, but I don't. How are you doing?"

"Fine, but I'm sure hungry."

"If you'll look in my bag over there you'll find some tea and some chocolate and those loaves of bread Bruce sent with me. I don't feel like eating, but help yourself. But I would like some tea if you don't mind brewing some."

"No problem," Andrew replied. He wondered how much he should eat since there was no way of knowing how long they would be there, so he ate less than he would have liked.

As they sipped their tea, Walker suggested, "Let's talk, I think it would get my mind off the way I feel and help pass the time."

"Sure," Andrew agreed, "You go first".

Walker began sharing some of his experiences, eventually telling much of the story of his life, interrupted only by times when he had to attend to his "bodily needs" and when Andrew had to gather more firewood. He shared how he:

-was born in Xenia, Illinois,

-became a businessman at age nineteen when he contracted with a railroad to furnish ties,

-tried teaching for awhile,

-worked as a surveyor for the St. Paul and Duluth railroad,

-invested his savings in pine forest land,

-had a mill at Crookston, Minnesota, and

-then came east to the Crow Wing and Leech Lakes area.

He ended his biographical discourse by saying, "And now, Andrew, I'm fifty-eight years old, old enough to know better than to get us into a mess like this; but now I want to know all about you, so start at the beginning."

"Well, it doesn't take long to talk about fifteen years," Andrew paused and then asked, "What date is this?"

"I believe it's the seventeenth, why?"

"Would you believe today is my sixteenth birthday?"

"Well, congratulations, Andrew! This is a poor place for a celebration, but I'll guarantee you the best meal in town when we get there!"

"Thank you, Mr. Walker; so, back to my life's story."

Andrew went on to tell about:

-his father and mother who were Swedish immigrants,

-how his father had worked for the Walker Lumber Company for nine years (Mr. Walker said he thought he remembered him),

-the farm and his father's encounter with the bull,

-how he was able to come to logging camp because of his dad's injury,

-going to a one-room country school, and

-he even remembered to mention his sister, Louise.

Mr. Walker interrupted frequently with questions, showing genuine interest.

When they had finished sharing about their lives to that date, Walker suggested, "Now, let's talk about our dreams for the future." He began unfolding his personal visions by saying, "I think my logging and lumbering business is just about where I want it to be. With a good man, Frank Akeley, as my partner, I feel less pressure. I'm making money, good money, more than enough to supply my needs and the needs of my family. I'd like to do some special things for others."

Walker paused, so Andrew asked, "Like what?"

"Well, a couple of things come to mind. I'd like to build, or at least help build, a big, beautiful Methodist church, an inspirational place to worship, a way of thanking God for His goodness to me, probably in Minneapolis. And then, promise you won't laugh?"

"I promise."

"I've always been interested in art. I know that sounds strange for an old lumberjack like me, but I'd like to build an art gallery, a place where people of all walks of life could come and see some of the beautiful artistic creations of this world."

As Walker spelled out his dreams, Andrew marveled how the man was thinking in dimensions he could scarcely comprehend. He just said, "Wow," at least a dozen times.

Finally, Walker stopped sharing and turned to Andrew. "Let's hear about your dreams. What do you want to become?"

Andrew responded, "I guess my dreams are pretty small compared to yours. I just want to have a farm next to my folks. In fact, with the money I've earned this winter, we'll buy the last 40 acres we need for two 160 acre farms. Then each winter, of course, I want to work for you."

"I'm flattered, Andrew, but what about an education?"

"My folks want me to attend high school in Little Falls, but I can't see why I need more education. My dreams are to be a farmer and a lumberjack. I have plenty of education for that now."

"Andrew, I just want to say that I'm really impressed with you. You have a lot of ability, just don't limit yourself. I would encourage you not only to finish high school but also to think about college. There's a fine new Methodist university in St. Paul called Hamline. I could help you get in there after you finish high school."

"But the cost—"

"I'd give you employment in the cities while you went to school. Who knows where an education might lead you, what doors it might open?"

Andrew was overwhelmed. "I'm sure grateful to you, Mr. Walker. I promise to think about it, honest."

So the day passed. Conversation eased the mind, and chunks of Bruce's good bread helped ease the pangs in Andrew's stomach. Night came, but with no change in the weather. Andrew wondered what they would talk about the next day. Before dark, Andrew again laid up a healthy supply of firewood as the travelers faced another night in the wilderness. They had begun the day as fellow travelers, but ended it as friends, good friends.

Sometime around midnight, Andrew awakened and tended the fire. He became aware of a marked change in the weather. The wind had gone down and it had stopped snowing. He looked up at the sky and saw stars. The storm was over.

In the morning Mr. Walker announced that his stomach felt much better, "But I'm still pretty achy all over," he added.

"Do you feel good enough to try going on to Akeley?" Andrew asked.

"Oh, I think so. Let's give it a try."

Dan and Dolly walked through snowdrifts that were sometimes belly-deep, but at least the trail was unmistakable and in a couple of hours Akeley was in sight.

First stop was the cookshack where Walker ordered, "Give this young man anything and everything he wants to eat!"

Walker, himself, ate very little, not wanting to upset his tender stomach, and then announced that he was heading for a hot bath and bed. "I'm going to start up the fire and put on as many covers as it takes to get warm clean to the bone!"

Before leaving Andrew in the care of the cook, George Salisbury, Walker turned to Andrew and ordered, "Put out your hand."

As Andrew responded, he felt something smooth and hard being deposited in his palm. Walker then spoke, "I want you to know, my young friend, that I thank you for what you did for me from the bottom of my heart, and I'm going to continue keeping my eye on you."

When Andrew opened his hand, he saw a gleaming twenty-dollar goldpiece!

CHAPTER XIII

HEADING HOME

George Salisbury was not like Bruce. In fact, the two men were about as different as two men in the same occupation can be. Bruce was a huge man with a black, bushy beard. George was small, wiry, and clean shaven, except for a tiny mustache about as wide as his narrow, long, boney nose. Bruce was cheerful; he often broke into song. George was grumpy. Andrew couldn't recall ever hearing Bruce swear. George would use cuss words in nearly every sentence he uttered, some of them Andrew had never heard before and he wasn't even sure what a few of them meant. What the men had in common was that they were both good cooks and easy to work for. Although not a lot of fun to be around, George made it very clear what he expected, and if Andrew performed accordingly everything went well.

The snow deposited by the spring storm melted in about ten days, and the lakes and streams began to open along their shorelines. While waiting for the ice to go out, Andrew helped George and the other members of the cooking crew prepare meals for the workers in the Akeley mill. As the breakup appeared imminent, Andrew and George began organizing provisions for the log drive. Just as soon as there was enough open water along the shore, the wannigan was pushed into the lake so that the bottom boards could swell up with moisture, closing the cracks and making the boat water-tight. It was not a small boat. It was a flat-bottomed scow about thirty feet long with an enclosed cabin almost the full length. There was a set of bunks in one corner and a kerosene stove to cook on. "Wouldn't dare use a wood stove," Salisbury observed, "Probably burn the blankety-blank boat down!" A work table, cooking utensils, food, and other provisions filled nearly half the cabin. The remainder of the space was mostly occupied by a long dining table with a bench on each side. It would serve about a dozen lumberjacks at a setting. Because of the limited facilities, the menu would not be as complete as back in logging camp. There would be a lot of soups and stews and sandwiches. A small oven sat on top of the kerosene stove. It did a fine job of baking bread and cookies, but because it was so small, it was kept going most of the time between meals.

The ice went out, totally, on the thirteenth of April. The next day, a small steamboat called "the Seagull" pulled the wannigan through the chain of lakes (there were eleven of them) which serve as the source of the Crow Wing River. During the winter, logs cut within a half-dozen miles or so of Akeley were hauled in by team to the mill to be sawed into lumber. Those cut farther away and near the lower three lakes or the river itself were stockpiled on the shore or on the ice itself. They were to be delivered to the big Weyerhauser mill in Little Falls. All of them couldn't be parked on the ice, there just wasn't enough room and it would have caused one gigantic logjam. Because the Crow Wing is such a narrow river in spots, there were a great number of logjams along the way as it was. Since there were other operators along the river, the logs were "branded" on their ends with the trademarks of the companies they belonged to so that they could be sorted out to their rightful owners once they reached their destinations on the Crow Wing and the Mississippi.

Twenty 'jacks rode along on the Wannigan and the Seagull. It would be their job to keep the logs moving down stream and to break up jams wherever they occurred. Each man was armed with a pike pole with a metal hook on one end. This was their only tool for pulling apart the pile-ups. It was dangerous work. Most jams could not be untangled without walking on the logs, and they were plenty slippery and unstable. The lumberjacks' hobnail boots helped, but it was a rare day that a man did not get a cold bath. When the boats reached the farthest Crow Wing Lake (called First Crow Wing), the Seagull pulled the wannigan into the current of the river and then cut it loose, returning to Akeley where it would pick up another wannigan which would serve food to the lumberjacks making the second drive with the logs that had been stored along the banks. This wannigan was without power of its own; the current carried it along at a plenty fast clip. In fact, it was important to have a couple of good anchors aboard to keep the boat from slamming into the logjams as they occurred along the way.

As the wannigan followed the logs downstream, Andrew was treated to a good view of the countryside. The trip was fascinating, but the scenery was marred by the barren landscape. Only the smaller trees and the hardwoods escaped the loggers axe and saw. In some places, huge, waist-high stumps covered the landscape as far as the eye could see. Andrew couldn't help but wonder if the forests would ever grow back. It was the first time he had thought seriously about what the logging industry might be doing to the environment.

Andrew noticed that every few miles there were wing dams made of timbers. These were tied open along the banks. George explained that after the spring run-off had subsided, the level of the river would go down markedly. Only by closing these dams could the water level be raised high enough to float the logs that remained upstream. He said that teams of horses or oxen were used to pull the wing dams open against the current.

Eight days after leaving First Crow Wing Lake, the wannigan reached the confluence of the Crow Wing and Mississippi Rivers. Here, the Walker Company logs mixed with an even larger number of logs from up the Mississippi. It became increasingly apparent how important it was that each company mark their own logs.

Two days later the crew arrived at Little Falls, the final destination for the wannigan. The river town was bustling with activity, most of it having to do with the lumber industry. On the river itself, logs belonging to each company were sorted and then grouped together by chains of logs, tied end to end. These were called "booms". Each raft of logs would then be guided to the proper mill along the river, some going as far as Minneapolis. The largest logging operation at Little Falls was that of the Weyerhauser Company to which Walker had sold the logs.

Andrew and his boss reported to the Walker Company office for work assignments. George was given new cooking responsibilities. Andrew was told he would be used to run errands for a few days but would then be free to return home on Saturday. He found it difficult to believe his tour of duty was nearly over. It had been a great adventure, but now he could hardly wait to see his family and friends and get back on the farm. He kept looking for some neighbor from home who could get word to his father to pick him up on Saturday. Thursday morning he saw Gust Kuhlander driving a wagon down mainstreet. Andrew had to run after him to get his attention, but was rewarded by his neighbor's assurance that he would stop by to see his folks on his way home that afternoon.

Andrew slept little Friday night. "It's funny," he thought to himself. "I'm as excited about going home as I was about leaving for logging camp last fall."

First thing Saturday morning Andrew walked to the Walker office, down by the river, and waited for it to open so he could be paid. He was just standing there on the shore of the Mississippi, taking in the view across the river, his hands in his pockets, minding his own business, when someone suddenly gave him a hard push from behind, sending him sprawling into the water! It was not deep, but deep enough to get very wet.

"What the —?" he muttered, scrambling to his feet and turning around; it was Tom!

"Hi," Tom said innocently, waving one hand.

"Tom Johnson, you idiot!" Andrew screamed, leaping out of the water and burying his right shoulder into his friend's stomach. "Revenge time!" he yelped, taking Tom to the ground. Andrew really hadn't wanted to test his added weight and strength against his older buddy quite this soon, but being pushed into the water from behind was just too much, and he had to do his best to get even. The boys rolled over and over, with no one staying on top long enough to be in control. Andrew was really fired up and the match was far different from previous brawls when Tom could do pretty much as he wanted with his younger friend; now Andrew was able to break every hold. With his added weight and strength he wore Tom down, finally putting him flat on his back. Andrew lay across his chest with his left arm under Tom's neck and his right arm under the opposite armpit. He was able to lock his arms together behind Tom's back and then squeezed for all he was worth. Tom bucked and kicked and squirmed but he only succeeded in using up more energy and wearing himself out. "Oh, revenge is sweet!" Andrew sneered, "and you said I'd never be able to handle you, ha!"

Tom finally stopped struggling and just lay there with his knees up and feet resting flat on the ground as he wheezed back, "I'll never give up, you can hold me here forever, but I'll never give up!"

"Fine with me," Andrew answered laughing, "I've got all day." But Andrew was not satisfied just being on top, so he made a surprise move. Without losing control, he pulled his right arm out from underneath Tom and slipped it down under Tom's knees. He quickly drew the knees up against Tom's chest, tight, until he could reach back under and grab his own wrist, holding Tom in a cradle. Slowly and carefully, Andrew began getting to his feet, lifting Tom off the ground.

"What are you doing?" Tom gasped.

Andrew didn't answer; he just walked out into the river until the water was well over his knees and then dropped Tom!

As he helped his drenched opponent to his feet, Andrew asked, "Even?"

"Even," Tom acknowledged, forcing a grin.

"Where did you come from?" Andrew asked, still surprised to see his buddy from logging camp.

"Got in last night, they put me on the second wannigan out of Akeley."

The boys headed for a company storeroom where they changed to dry clothing and caught up on all that had happened since last they were together.

One of the first questions Andrew had for Tom was, "Did the camp fill its quota of logs?"

"You bet," Tom replied, "and about 250 logs over to boot."

Andrew then related the adventures of his journey with Mr. Walker, ending the story by showing Tom his twenty-dollar goldpiece.

Tom just shook his head and said, "I just can't believe how lucky you are."

About that time the company office opened and the boys joined a line of workers waiting to be paid. Andrew received $182, including the three day bonus for making quota. Mr. Walker had left directions that he share in the extra pay. Andrew was told that his father would be receiving a check for the use of the horses. He felt very rich.

"I should buy some Christmas gifts for the family. Want to go along?" Andrew asked Tom.

"Sure, I need to do the same."

The boys headed for the retail store area of town. Except for Andrew's brief stay in Walker and Akeley, it was his first opportunity to spend any money.

Upon returning to the company office, Andrew was surprised to find his mother, father, and sister all waiting for him. It was everything a family reunion should be; laughter, tears, hugging, and everyone trying to talk at the same time. Tom stood awkwardly by until Andrew's mother finally asked, "Aren't you going to introduce your friend?"

"Sorry. This is Tom Johnson, the friend I talked about in my letters."

After introductions were complete, Andrew's mother addressed Tom, "I understand you live down by St. Cloud. That really isn't far from Upsala. You'll have to come visit us this summer."

"Thank you," Tom answered. "Andrew and I have talked about that. If it's all right with you folks, I thought I'd come up for the 4th of July."

"That would be fine," she assured him.

As Mr. Anderson loaded Andrew's things into the surrey, the boys began their good-byes. Tom pulled Andrew to one side and whispered, "You never told me your sister was so good looking! How could anyone as ugly as you have such a pretty sister?"

Andrew had never really thought about his sister as being particularly beautiful, but retorted, "Boy, if you think she's beautiful, the women you know must be some real cows!"

Tom gave Andrew a push, but then pulled him back and continued in a low voice, "Fourth of July, heck! I can't wait that long to get to know her. Look for me in a couple of weeks!"

When the family reached home, Andrew sat them all down around the kitchen table and presented the gifts he had purchased earlier that day. "Merry Christmas!" he said laughingly, and gave each a package. For Louise and his mother he had purchased four different kinds of material so that each could have two dresses made. For his father, he had a pearl handled pocket knife and a new logger's cap to replace the battered one the father had placed on Andrew's head as he started the trip north.

"You must have spent all the money you earned," Mr. Anderson teased.

"Well, not really," Andrew replied, and counted out $173, laying one bill at a time in the center of the table. "I'll bet that just about equals anything you ever brought home, Pa" Andrew boasted.

"Almost, I'll have to admit, but it seems to me that after buying gifts last year I had about $191 left.

"Huh, I can beat that, 'cause I'm not through counting!" and with a dramatic wave of his hand, Andrew placed the twenty-dollar goldpiece on top of the stack of bills.

His mother said, "Oh my goodness!"

Louise said, "Ohhhh!"

His father said, "Well, I'll be a monkey's uncle!"

There were no farm chores done the rest of the day until milking time, as Andrew told all of the stories contained in this book.

PHOTO SECTION

Courtesy Minnesota Historical Society

Lunch in the woods, served by a "swing dingle" like Andrew drove.

Courtesy Minnesota Historical Society

Spring comes to a turn-of-the-century logging camp.

Courtesy Minnesota Historical Society

Bunkhouse. Note the clothes drying over the stove.

Courtesy Minnesota Historical Society

Logging camp dining hall, much like the one where Andrew and Tom worked.

Courtesy Cass County Historical Society

Steam hauler pulling loads of logs around the turn of the century, a few years after Andrew's first year in the woods.

Photo courtesy Ed Morey

Logs destined for the Crow Wing River—and then to help build America (early 1900's).

Logging railroad, Cuba Hill area. This may be the trail now used as the Sucker Bay road which passes the Cuba Hill forestry tower.

Courtesy Minnesota Historical Society

Corduroy logging road. When spring came too early, this was an effective way of transporting logs over swampy areas.

SAWMILL OPERATIONS AT MOTLEY ON THE CROW WING RIVER

All photos courtesy Ed Morey

The barrels on top of the building are filled with water for use in case of fire.

Courtesy Minnesota Historical Society

Chief Flat Mouth, the younger. His refusal to support "Old Bug" helped spell doom for the Indian revolt. Andrew loved listening to his stories.

Eshke-bog-e-coshe (Flat-Mouth, the elder) of Leech Lake. Bust by Francis Vincenti. Location: Senate wing, third floor, east. (One of only three Indian statues in the U.S. Capitol). He was the father of the Flat Mouth Andrew met.

Courtesy Minnesota Historical Society

Troops arrive in Walker, October, 1898. The year Andrew was first a lumberjack.

Courtesy Minnesota Historical Society

A cabin and garden belonging to "Old Chief Bug"—the setting for the last Indian-White war.

Courtesy Minnesota Historical Society

This is believed to be the first photograph of the village of Walker—taken about 1896, two years before the last White-Indian war.

Courtesy Minnesota Historical Society

The village of Cass Lake was worried about the Indian uprising, too. This fortification was still standing two or three years after the battle when this picture was taken.

Courtesy Minnesota Historical Society

Old Chief Pugona-geshig (on the left), nicknamed "Old Bug" (the Ojibway "P" is pronounced "B"). He was the focal point of the last White-Indian war in our country.